D1615818

FUNDAMENTAL PRINCIPLES OF SOL-GEL TECHNOLOGY

R. W. JONES

The Institute of Metals
1989

Book 475   CHEMISTRY

published by

The Institute of Metals
1 Carlton House Terrace
London SW1Y 5DB

distributed in North America by

North American Publications Center
Old Post Road, Brookfield
VT 05036, USA

© 1989 THE INSTITUTE OF METALS

All rights reserved

British Library Cataloguing in Publication Data

Jones, R.W. (Ronald William), 1945-
Fundamental principles of sol-gel technology.
1. Glass ceramics
I. Title
666'.15
ISBN 0-901462-69-1

Library of Congress Cataloging in Publication Data

Jones, R.W. (Ronald William), 1945-
Fundamental principles of sol-gel technology / R.W. Jones.
p. cm.

1. Colloids. 2. Ceramic materials. 3. Glass. I. Title
QD549.J655 1989
541.3'45—dc20

ISBN 0-901462-69-1

Cover design: Jenny Liddle

Text processing by P i c A Publishing Services from original material provided by the author

Printed and bound in Great Britain by M & A Thomson Litho Ltd, East Kilbride, Scotland

CONTENTS

1 HISTORY   1

2 INTRODUCTION   3

3 THE COLLOIDAL STATE   5

3.1 CATEGORIES OF COLLOIDS

3.2 PHYSICAL CHEMISTRY OF COLLOIDS
    3.2.1 Permanent dipoles
    3.2.2 Dipole induced in a non-polar molecule by a permanent dipole
    3.2.3 Resonant induction of dipoles

3.3 ELECTRIC DOUBLE LAYER
    3.3.1 Development of surface charge
    3.3.2 The structure of the double layer

3.4 COLLOID STABILITY
    3.4.1 Colloid stability and agitation
    3.4.2 Peptization

3.5 EPILOGUE

4 SOL-GEL TRANSITION   33

4.1 INTRODUCTION

4.2 THE FORMATION OF GELS

5 FORMATION OF SOLS   40

5.1 BALL OR JAR MILLS

6 SOL-GEL TRANSITION IN ALKOXIDES   49

6.1 ALKOXIDE SYNTHESIS

6.2 ALKOXIDE SALT METHODS

6.3 MULTICOMPONENT AQUO-SOLS CONVERTED TO GELS

6.4 OTHER ROUTES TO GELS

6.5 MIXED SYSTEMS

6.6 CHELATION

7 THE GEL TO GLASS OR CERAMIC TRANSITION   67

7.1 DRYING

7.2 DRYING CONTROL ADDITIVES (DCCA)

7.3 HYPERCRITICAL DRYING

7.4 DENSIFICATION AND SINTERING

8 APPLICATION, RECIPES AND REVIEWS   81

8.1 SOL GEL COATINGS

8.2 ORGANICALLY MODIFIED SOLS

8.3 CERAMICS FROM SOL-GEL
    8.3.1 Densification

8.4 ELECTROCERAMICS

8.5 GLASSES AND GLASS-CERAMICS VIA SOL-GEL

8.6 MICROSPHERES

8.7 SOL-GEL FABRICATION OF FIBRES

8.8 SOL-GEL IN THE PRODUCTION OF REFRACTORIES

8.9 SOL-GEL FOR CATALYSIS

8.10 MATERIALS FOR ELECTRONICS

8.11 BIOMATERIALS AND SOL-GEL TECHNOLOGY

8.12 COMPOSITE MATERIALS

REFERENCES   **127**

To Antonia, Michael and Jenny

who thought a book about Sol-Jellies would be a very good idea!

1   HISTORY

The preparation of inorganic material via chemical routes is not new but currently it is treated as modern technology. As with many scientific developments, one can normally delve into the past and find some reference to the discovery but unless there is a day to day requirement at that time for the technology, it generally lapses into the realm of scientific curiosity.

> Sol-gel is a simple technology in principle but has required considerable effort to become of practical use. Many people have experienced the phenomenon in their basic qualitative chemical analysis at school when they have developed gelatinous precipitates of metal hydroxides. The fact that one can suspend a solid in a liquid, then remove the liquid and finally densify the solid, can be of great help in the preparation of glasses and ceramics. Sol-gel enables materials to be mixed on a molecular level then brought out of solution either as a colloidal gel or a polymerized macromolecular network whilst still retaining the solvent. This solvent can then be dessicated off leaving a solid with a high level of fine porosity. So high is the porosity that these solids possess very high surface area and therefore surface free energy and it is this property that enables the solid to be sintered and densified at much lower temperatures than one would expect in the normal processing of the same material. For instance, a temperature of 1750°C is required to fuse silica, SiO_2, yet a dense solid can be prepared from a gel at 1100°C.

Some of the earliest work on sol-gel appears to have been reported by Blodgett (1) in 1935. Forty five years ago, Geffcken and Berger (2), who were working for Jenaer Glaswerk Schott at the time in Jena, patented oxide coatings that they had prepared by the sol-gel method. This patent was published in 1943 and during the second World War silica was produced as thin films from colloidal solutions of silicic acid but this was quickly overtaken by vacuum deposition techniques. By 1953 the concept had been commercialized but it was not until 1959 that volume production of coated rear view mirrors for the automotive industry began. These glass mirrors were coated with (TiO_2-SiO_2-TiO_2).

By 1969 Schroeder (3) had produced a classic publication on oxide layers deposited from organic solutions. He described dip coating, spinning and spray coating. He also covered a range of single and multi-component oxides. The publication contains many practical aspects of coating including surface preparation and multi-layer coating. There are now numerous patents but some of the early work is still of great practical use (4, 5, 6).

Apart from coatings, colloidal forms of silica and alumina have been used for many years as bonds in refractory materials. These colloids are generally aqueous. They are mixed in with the refractory aggregates and help form a bond between the coarser particles during sintering.

Another classic publication by Stoeber, Fink and Bohn (7) in 1968 described a development which eventually became known as the Stoeber process for the controlled growth of monodisperse silica microspheres. This was achieved by base catalysis of alkyl silicates and at the time the spheres were considered to only be of academic interest, but this is now a very useful way of producing precursor powders for the manufacture of high performance ceramics.

I have cited examples above of sol-gel technology but these are rather isolated and one can argue that the true development of this field has not taken place until the last decade. There are now several commercialized processes and an increasing amount of research in this field. Advanced applications are now rapidly evolving including:

- synthesis of superconductors
- coatings on optical memory discs
- large mirrors for laser weapons in space

This work is likely to gain momentum towards the turn of the century and to an extent this may displace some of the more conventional glass and ceramic technology we are familiar with today.

2 INTRODUCTION

The term sol-gel is now frequently used to describe many aspects of the chemical synthesis and processing of inorganic materials such as ceramic powders and glasses. Unfortunately, the term is used very broadly and covers a wide range of colloidal and macromolecular systems and therefore requires some degree of subclassification. In general, this can be considered in terms of the system which develops from the sol. We can accept that in all cases the sol is a system which allows chemical species to become stable in solution but on a level where the species have either acquired a particle size or molecular weight whereby the sol is only just stable and if the conditions of solvation or suspension are changed slightly this can lead to destabilization of the sol as:

- precipitation of the sol species as aggregated particles
- precipitation of unaggregated particles
- formation of a homogeneous gel

As to which of the above three states are achieved depends upon fairly complex chemistry but can be generalized as follows:

- the starting point for sol formation
- the nature of the solvent
- the method used to catalyse the colloid precipitation or gel formation

To a large extent much can be understood about sol-gel technology through an understanding of the physical chemistry of colloids and it is this subject that will be dealt with in the first chapter. Subsequent chapters will help to differentiate between true colloidal sols and the more recent macromolecular gels usually associated with the polycondensation of metal alkoxides. It will be illustrated how each system leads to particular forms of materials such as powders, spherical particles, coatings, fibres and monoliths. It is possible to achieve many technologically useful shapes and forms of materials and to control composition and purity much more easily than by conventional processing. For example, near net shapes can be formed in high purity silica which can be used to construct laser mirrors for use in space. Coatings can be applied to architectural glass to enhance solar reflection. Fine ceramic powders can be formed which are uniform in size and sinter to high density at temperatures much lower than by conventional powder processing. Sols can be filled with other ceramic powders and used as a binding phase to enable good dense bodies to be obtained with minimal shrinkage. Glasses can be prepared at temperatures as low as one third of the fusion temperature thereby avoiding the problems associated with high temperature fusion:

- crucible contamination
- expensive furnacing
- difficult homogenisation of the melt

Very thin coatings can be applied to almost any geometry of substrate and these coatings can be multi-layered to build up complex surfaces exhibiting nonlinear optical effects or surfaces not prone to laser damage. Fibres can be drawn from partially hydrolysed gels and subsequently densified by heat treatment. Because it is possible to chemically dope the starting material, this technique can be used to make optical fibres.

The above examples are but a few instances of the usefulness of sol-gel technology. There are many other applications which will be described in more detail later in the text. It is the aim of the next two chapters to develop the basic theory of colloids and gels and, as such, the text relies on the knowledge of physical chemistry. Where theory has been necessary, this has been included but references are also given to other sources to enable an indepth study to be made of individual topics.

3 THE COLLOIDAL STATE

The science of colloids is a very well reported subject. This has arisen because colloids exist in many forms and are technologically important.

Colloids are to be found in such every day materials as:

•   food (flour as a food thickening agent, souffles, mousses)
•   cosmetics (many kinds of hair shampoo, shaving foams, etc.)
•   aerosols
•   paint (modern paints generally are colloidal suspensions and require no stirring to bring the pigment into the body of the paint)
•   coloured glass (colloidal gold was a common form of red glass in stained glass windows)
•   natural silica gels (opals and agate)

It is obvious then that colloids can exist as systems comprised of the following:

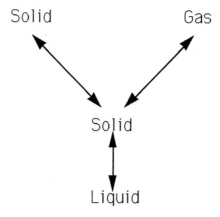

Figure 1

and another combination liquid/liquid which is more commonly known as emulsion.

In this monograph, it is not possible to give a comprehensive discourse on colloid chemistry. However, I will try to include the necessary currently accepted theory or to give reference to other texts which may deal in more detail with the topic.

To use the term "physical chemistry" throughout this monograph is not intended to be a lack of distinction between the importance of either physics or chemistry but rather to highlight that the discipline where both sciences have formed a hybrid is the only field that has made a specific study of the nature and behaviour of colloids.

We begin our discourse on the colloidal state by considering the behaviour of macromolecules in a second phase which will generally be a liquid but which can also be another solid or a gaseous phase. Molecules and atoms can sometimes form conglomerates under the influence of intermolecular forces. If these conglomerates remain stable, they are referred to as *colloids*. The state we are most interested in is the colloidal suspension of a solid in a liquid and this is called a *sol*.

If we examine the thermodynamics of macromolecules dissolving in solution, we find that it is the entropy change which is important. The solutions formed by macromolecules are referred to as non-ideal. This is because molecules displace large amounts of solvent instead of simply dissolving without disturbing the solvent. There is a large excluded volume which means that a molecule is unable to swim freely through the solution because it is not able to occupy regions where there are similar molecules.

As well as an entropy change, there may also be appreciable enthalpy of solution if there is an interaction between the solvent molecules and functional groups on the macromolecule.

The suspension of the sol particles in solvent is best thought of as a dynamic process in equilibrium. Obviously if the density of the solid is higher than that of the liquid, the particles will want to sediment out of solution. Several theoretical analyses have been developed to describe rate of sedimentation. These analyses require approximations which are not always valid assumptions to be made in complex macromolecular systems. Firstly, most analyses consider the sedimentation of uncharged particles which are of uniform mass, m. This concept of a "monodisperse" sol is often far from reality. Stoke's Law is used to describe the frictional effects of the liquid against the descending particle under the influence of gravity and here Stoke's Law assumes:

1   The particles are spherical.

2   The solution or suspension is very dilute.

The Colloidal State

3   The molecules in the liquid medium are considered to be very small compared to the colloidal particle and can thus be treated as a continuum.

Again such assumptions have limited value in colloidal science. The sedimentation process will be opposed by thermal agitation. The molecules in suspension are in continual collision with solvent molecules and the walls of the vessel and display a random motion through the liquid known as Brownian motion. Thus, it is possible to establish a sedimentation - diffusion equilibrium. Use is made of this equilibrium to measure molecular mass. Again problems arise with this technique if the suspension is polydisperse (i.e., consists of particles of different sizes).

Now to discuss a little more about the nature of the colloidal particle itself. Generally, we consider a colloidal particle to be 500 nm or less. These aggregates of atoms or macromolecules cannot be seen with a normal optical microscope since their maximum size is equal to the wavelength of light. More indirect techniques are needed to study colloids which include:

- light scattering
- sedimentation analysis
- osmosis

Light scattering is a most useful tool in colloid technology. The kind of information that can be obtained by light scattering is:

- size of the colloidal particle
- shape of the colloidal particle
- interaction between particles

The advantages that light scattering has over many other techniques are:

- The technique is absolute and requires no calibration.
- Measurements are almost instantaneous.
- A large proportion of particles are sampled at once.
- The structure of the system is not disturbed.

The parameters which give the information about particle size and geometry are the intensity, polarization and angular distribution of the scattered light. Modern techniques employ the use of lasers.

3.1 CATEGORIES OF COLLOIDS

There are many examples of colloids but it is possible to separate them into several general categories.

Monodisperse Colloidal Sols

Here all the particles are identical in size and shape. [There is some difficulty encountered in the preparation of such sols since there is a tendency for the growing particles to aggregate. However, under controlled conditions, particularly by seeding supersaturated solutions it is possible to achieve this state]. Monodisperse sols are nowadays appreciated as a route to production of high quality ceramic powders and several monodisperse systems have been synthesized from hydrous metal oxides such as zirconia, titania and alumina. Silica sols are also well known as being capable of being developed in a monodisperse form.

Polydisperse Colloidal Sols

This is more typical of colloidal systems where the particles have a definite size and/or shape distribution. In such systems, we cannot give a strict meaning to the molecular weight or particle size. Instead, we need to consider the system statistically and therefore define average quantities. The significance of these quantities depends on the relative contribution that each species makes to the phenomenon used as the measurement technique. When colligative properties are examined, e.g. osmosis, we find such properties depend on the number of particles present so the measurements give a number average molecular weight. The problem is that larger particles often make a larger contribution to the property being measured and thus a weighted average must be established. If the contribution of each particle is proportional to its mass (e.g. light scattering) then a mass average molecular weight can be defined.

If M_m is the mass average and M_n is the number average in a typical polydispersed system then

$$M_m > M_n$$

but in a monodispersed system

$$M_m = M_n$$

Many colloids occurring in nature exhibit polydispersity. Both monodisperse and polydisperse sols which are true colloids are thermodynamically unstable.

The Colloidal State

The particles have such a high surface free energy that after the particles grow and phase separate into a solid precipate, the particles cannot be reconstituted into the sol again. Thus the reaction is irreversible. Typically, these colloid sols tend to be based on aqueous systems which yield fine particles that can be used as ceramic precursors. As will be discussed later, the formation of colloidal particles is influenced very much by charge effects around the particles. This will be discussed under the section dealing with electrical double layers. Other important factors are:

- the dielectric constant of the solvent
- the presence of ions that can influence the charge interaction between the particle and solvent

Macromolecular Sols

These systems are different in several ways to colloidal sols. In this instance, the growth of the particle or mass is brought about by polymerizing and cross-linking between the polymer molecules. As the polymeric species grow and interconnect, eventually solvent is trapped within the system and the resultant phase becomes a semi-solid known as a gel. As will be seen in detail later, these systems are often derived from organo-metallic solutions in alcoholic solvents. It is this third category that has received the main focus of attention in the past decade and is commonly referred to as sol-gel technology. The basic difference between these macromolecular sols and conventional colloids is the fact that on gel formation these systems are often reversible. The gel, even in the dried form, can be redispersed back into the solvent.

A range of applications can be found for the use of colloids and macromolecular gels and there are areas where hybrids of each species exist. For instance, it is possible to begin with a polymerized system based upon organic starting materials and solvent but with careful catalysis and additions of electrolytes it is possible to convert the gel into a conventional colloid and finally precipitate the colloid, thus passing from a reversible to irreversible system. The main point to be understood here is the incredible versatility of CHEMICAL SYNTHESIS of inorganic materials allowing:

1   Control over composition

2   Molecular homogeneity

3   Many forms in which a phase can be prepared - powders, coatings, monoliths and fibres

4	Many forming techniques for producing materials as in (3) often requiring very low process temperatures

3.2 PHYSICAL CHEMISTRY OF COLLOIDS

Texts often refer to colloids as lyophilic (solvent attracting) or lyophobic (solvent repelling). I do not wish to define colloids in this way since it can lead to ambiguity. If we take the example of silica or alumina colloidal aqueous solutions, these would generally be referred to as lyophobic but there must be some degree of attraction there for the solvent or a colloidal dispersion would not exist in the first place. Likewise, lyophilic systems of macromolecules often have lyophobic functional groups attached to them.

It is better, therefore, to dispense with this simple classification and to study some of the more complex reactions that take place between the surface of the colloidal particle and the solvent molecules. The colloidal particle is essentially an unstable species. Thermodynamics tells us that any particle with a high specific surface area will have an associated high surface energy.

If we imagine the work needed to increase the surface of a particle by a small increment $d\sigma$ then the work is the product of the force resisting increase in area times the distance it moves. All materials exhibit a resistance which we define as the surface tension X. Then the increment of work dw is $X\,d\sigma$ i.e. $dw = X\,d\sigma$

A more rigorous argument is that the work of surface creation is additional to pressure/volume work and if we examine the Helmholtz function relating the change of state of a system to entropy, p-v work and surface creation, we can express this as follows:

$$dA = -S\,dT - p\,dV + X\,d\sigma$$

dA   is the Helmholtz function

S   is entropy

T   is temperature

p/V is pressure/volume

All systems try to lower dA thus a system attempts to lower its internal energy and increase entropy or disorder. The additional term in the expression above

The Colloidal State

indicates the desire to decrease surface energy. If you are familiar with chemical thermodynamics, you are probably more used to using the Gibb's function, G, or the free energy since we are often more interested in changes occurring at constant pressure and not constant volume; whereas, the Helmholtz function considers changes in a system occurring at constant temperature and volume. Here we consider doing work to increase the surface area and thus the function A sometimes known as the "work function" is more appropriate in the above argument.

From this, it should be apparent that thermodynamically the colloidal state should not exist at all yet experience tells us it is common. The stability of a colloid is therefore a kinetic one. The particles in the solvent are trying to collapse; to actually move towards one another and coalesce to reduce surface energy.

Imagine two spherical particles 200nm in diameter. Their volumes are $4/3 \pi r_1^3 = 4.189 \, r^3$ and surface area $4/3 \, r_1^2 = 12.57 \, r^2$. We will use a simple unit of square area in nm which is arbitrary but fine for calculation. We have Volume $= 4.189 \times 10^6$ and Area $= 12.57 \times 10^4$. So the total surface area of the two spheres is $= 25.14 \times 10^4$. If we now combine the spheres to make one larger sphere, we need to calculate the new radius for the combined volumes. Original spheres were r_1 and new sphere is r^2

$2 \times 4.189 \times 10^6$ $\qquad = 4.189 \, r_2^3$

so r_2 $\qquad = 126.99$

So the new area of sphere r_2 is

$12.57 \times (125.99)$ $\qquad = 19.95 \times 10^4$

Thus, the area of a single sphere is much smaller than the two separate spheres for the same volume of material. This then provides a strong driving force for the particles to combine. This behaviour is common in many systems and is the driving force for the sintering of ceramic grains. Furthermore, there is also a mechanism whereby this can happen. The particles, when they pass close enough, will experience a force of attraction known as van der Waal's force. This is a very fundamental force that exists between all particles, atoms and molecules. The force is weak unless the atoms get close to one another. Forces that we are more familiar with such as electrostatic attraction obey an inverse square law such that the force varies with the reciprocal of the square of the distance of separation.

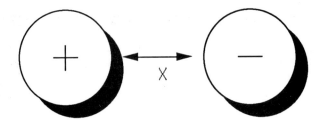

Figure 2

but van der Waal's force obeys a higher power law

van der Waal's force $\propto 1/x^6$

hence, the force drops away rapidly for quite small separations.

van der Waal's forces arise because of the permanent or induced polarisation in adjacent atoms or molecules even though the normal valence requirements in the molecules are satisfied.

The examples are given here.

3.2.1 PERMANENT DIPOLES

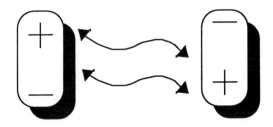

Figure 3

Molecules with permanent dipoles can orient in such a way as to produce attractive forces.

Attractive orientations correspond to a lower energy state than repulsive ones;

The Colloidal State

hence, in a fluid the net average orientations cause attraction.

3.2.2 DIPOLE INDUCED IN A NON-POLAR MOLECULE BY A PERMANENT DIPOLE

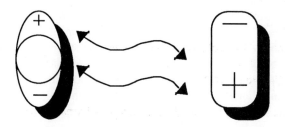

Figure 4

Here the electron cloud around the non-polar molecule is distorted and forms an induced dipole.

The strength of this interaction depends on the dipole moment of the polar molecule and the polarizability of the second molecule.

3.2.3 RESONANT INDUCTION OF DIPOLES

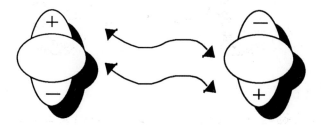

Figure 5

If the electron cloud in one molecule resonates, it can induce a dipole in an adjacent electron cloud leading to attraction. This induced dipole - dipole interaction is sometimes called the London attraction or dispersion interaction. The strength of this depends on the polarizability of each molecule or the looseness with which each electron cloud is held to the nucleus. This type of attraction also applies to polar molecules since they possess "instantaneous" dipoles. Thus, the total attractive force between molecules is caused by the sum of all three mechanisms described.

This argument has only considered dipoles but molecules exist with more complex electron distributions such as quadrupoles and higher. There are, thus, many more interactions but all lead to a net attractive force with a dependence on separation r proportional to $1/r^6$. Why then do colloids exist at all? Examining the surface area of a macroscopic centimetre cube of material, it has a surface area of $6cm^2$. Now divide this into cubes the size of a typical colloidal particle say 100nm cubes then the surface area increases to $600,000cm^2$. More finely divided at 10nm cubes the area increases to $6,000,000cm^2$. Thus, any effects connected with colloids are going to be surface dominated.

There are forces opposing the long range effects. The first can be considered as the physical stabilization of the particle. It is possible for the particle to develop a protective film at its surface often by reaction with the solvent. This protective film will not prevent particles repelling one another but if they touch it prevents penetration. There are many examples of such stabilization. A platinum sol will react in water to form $Pt-(OH)_3H_3$. This forms a layer or shell around the particle. The emulsification of fat by soap is another example of this behaviour. Here the hydrocarbon tails of the soap molecules penetrate the oil drop leaving the hydrophilic heads at the surface.

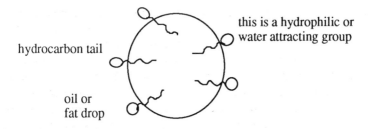

Figure 6

The hydrophilic heads then penetrate the hydrogen bond with water molecules. The above mechanisms are considered to be the physical stabilization of

The Colloidal State

colloids. However, there is another very important source of kinetic stabilization due to electrical charge on the particle surface and its association with surrounding ions in the solution. This effect is known as the electrical double layer and is now dealt with in quite some detail.

3.3 ELECTRIC DOUBLE LAYER

The surface of the colloid particle is capable of acquiring a charge when it is in a polar solution; for instance, water. This surface charge will have a strong influence on the surrounding ions particularly those of opposite charge known as the counter-ions. The counter-ions are attracted towards the colloid particle and likewise ions of the same sign, co-ions, drift away. This migration is continually disturbed by thermal agitation in the solution. Eventually an equilibrium situation is established in which a diffuse layer of counter-ions forms around the surface of the colloid. Interspersed with counter-ions are some co-ions. This complex atmosphere is, in effect, an electrical "double layer".

The first question we need to address is how does the colloid particle develop a surface charge in the first place? Later we examine the structure of the double layer and the size of the electrical potentials experienced near the charged surface.

3.3.1 DEVELOPMENT OF SURFACE CHARGE

Ionisation

In this effect the surface charge is developed by a functional group on the surface dissociating and forming an ion. As to the equilibrium concentration of ionic species on the surface and the sign of the charge, this will be determined by the pH of the solution. In an acidic environment, the protons in the solution attach to basic groups and the overall charge on the particle is +ve; whereas, in a basic solution, protons are lost from the surface giving the colloid a net -ve charge. There must then be a point where there is no net charge on the surface at a certain pH value. This point is known as the isoelectric point. If we plot a property related to the sign and intensity of charge, such as a transport phenomenon, against pH we can determine the isoelectric point. This is shown for a system in the figure below.

Ion Adsorption

As the name implies, this is the charge development on the surface due to the preferential adsorption of particular ions from the solution. In solution, all ions

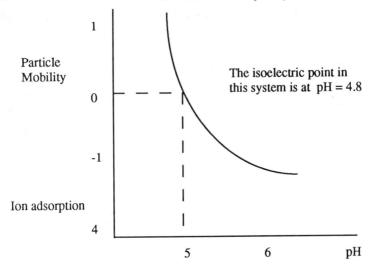

Figure 7

must be in a state of charge balance. That is to say, the numbers of charges on cations will equal the numbers on anions. Despite this equilibrium, the way these ions exist in solution can be very different. For instance, cations are generally surrounded by a hydrating layer in aqueous solution due to the strong field associated with cations. This tends to stabilize cations in the solution but the less hydrated smaller anions are more free to diffuse to the particle surface and are thus preferentially adsorbed. This would give the colloid particle a negative charge. If a surface is already charged by ionization, it is likely that it will absorb counter-ions and if this happens there can be a charge reversal. At this point, we need to think about the differentiation between adsorbed ions and the double layer. The double layer is a diffuse equilibrium which on the one hand is caused by a force of electrostatic attraction disrupted by thermal diffusion but eventually reaching equilibrium. Adsorbed ions form a tight, discrete shell around the colloid and are far from diffuse.

Ion Dissolution

When a colloidal particle is in contact with a solution, there will be some tendency for the surface to dissolve. If the particle is composed of two ionic species then it is likely that the anions and cations will dissolve at different rates.

This behaviour is common in metal oxide and hydroxide sols. Consider the example of a metal hydroxide in a solution with an excess of hydroxyl (OH^-) ions. In this case, the hydroxide particles become negatively charged, whereas, in an excess of H^+ ions the hydroxyls can dissolve more readily leaving a net positive

The Colloidal State

charge. Because OH⁻ and H⁺ have decided the nature of the charge on the colloid we refer to them as potential-determining ions.

3.3.2 THE STRUCTURE OF THE DOUBLE LAYER

It is reasonable to consider the double layer as two main regions. The inner region close to the particle is the tightly bound charged region surrounded by the more diffuse regions of counter-ions which establish an equilibrium distribution:

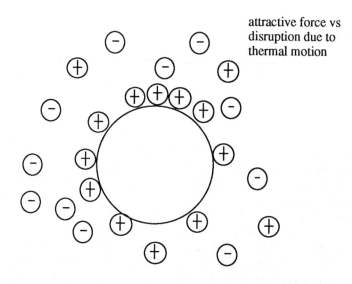

Figure 8

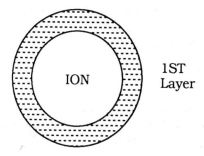

Figure 9

There have been several theoretical treatments made of the double layer with varying degrees of sophistication. All analyses require some level of simplification and certain assumptions to be made which may have limited correlation to real systems. Provided one does not wish to make accurate quantitative predictions from theory, the diffuse region can be described quite well using a theory of electroytes developed by two scientists, Debye and Huckel.

Essentially solutions of electrolytes deviate from the behaviour of "ideal solutions". At this stage, I do not intend to enter into the detailed thermodynamics of this statement other than to say that ideal solutions obey Raoult's Law:

$$P_A = \chi_A P_A$$

This law states that if we dissolve a species in a solvent, its vapour pressure will be proportional to the mole fraction of the material present in this solvent. Solutions deviate from ideal when the solute can interact with the solvent via forces other than van Der Waal's, e.g. electrostatic forces or associate in some way with the solvent, e.g. solvations of an ion.

Electrolytes are solutions in which electrostatic (charge) interactions dominate over weaker forces such as van Der Waal's and so a quantitative model which describes this behaviour should predict the deviations from ideality. We can treat an electrolyte as a simple version of a colloidal dispersion. The following can be said about the electrolyte:

- Oppositely charged ions attract each other.

- Because of this attraction, cations and anions are not evenly distributed throughout the solution.

- Whilst the overall solution is electrically neutral near any given ion, there is an excess of counter-ions.

- This leads to a diffuse averaged atmosphere of counter-ions around any one ion.

Thus, the energy of any central ion (or colloidal particle) is lowered as a result of the electrostatic interaction with its associated ionic atmosphere. *This can stabilize a colloidal particle.* What the Debye-Huckel theory sets out to achieve is to predict this effect quantitatively. Again, I see no reason to give a rigorous analysis of the quantitative modelling in this text as it is dealt with in many texts on physical chemistry. We are, however, concerned about the results and implications of this theory. Their model assumes that the average charge density at any point moving away from the surface of the charged ion arises from the competition between:

- the electrostatic attraction of the central ion for its counter-ions, and
- the disruptive effect of thermal motion in the solution.

We can define a substructure to the diffuse ionic atmosphere as follows: firstly, we have the ion (or charged colloid) with its surface charge and tightly bound to its surface is the first layer of counter-ions which are therefore immobilized.

In a colloid, the radius of the sphere that captures this tightly bound layer is called the:

- radius of shear

The radius of shear is an important factor in determining the mobility of the particles. If we measure the electric potential at the radius of shear and compare it to the electric potential somewhere distant in the solution, this quantity is called the:

- ζ (Zeta) potential

Hence, the combined inner bound shell and the diffuse atmosphere it attracts is called the electrical double layer. The resulting treatment performed by the Debye-Huckel theory models the diffuse layer as a series of concentric shells around the central ion. The radius of each shell r_D will depend on the charge type. This is related to the ionic strength of the shell. Let us consider the dissolution of a pair of monovalent ions, e.g. Na^+ Cl^-. The theory defines the ionic strength, I, as a dimensionless quantity dependent on the square of the charge number of the ions which is one in each case and the amount of the ion present determined by how much the NaCl salt dissociates. In dilute solution, the salt is totally dissociated; hence, the ionic concentration is the molality of the solution. Because the ionic strength is linearly dependent on the amount of any ion present and the square of its charge number, it is obvious that ions with higher charge number will generate higher ionic strength than lower charge number.

We have been considering the example of a salt, NaCl, comprised of monovalent ions and totally dissociated. If we use the expression:

$$I = 1/2(m_+ Z^2_+ + m_- Z^2_-)/M_o$$

(we can ignore the meaning of M_o here and assume it to be a normalizing constant that makes the ionic strength, I, dimensionless).

m_+ is the molality of the +ve ion and m_- the same for the anion

likewise Z_+ is the charge number of the cation ($Na^+ Z = 1$) and Z_- - 1 for the anion

$Z_+^2 = 1 - Z_-^2 = 1$

So $I = 1/2 \, (M_+ + M_-)/m_o$ but $M_+ = m_-$ therefore, $I = m/M_o =$ the total molality

Now let us consider a salt in solution where the charge numbers are different:

Salt MX_2

Here we have $M^{2+} + 2x^-$

A practical example is say $CaCl_2$. Again, complete dissociation in solution:

So $I = 1/2 \, (4m_+ + 2m_-)/m_o$

$= 3m/m_o$ because m+ = m and $m_- = 2m$

i.e. $I = 1/2 \, (4m + 2m)/m_o$

$= 3m/m_o$

So even though one can have two salts of the same concentration and totally dissociated the ionic strength of the salt with the higher charge numbers is greater. A simple table can be drawn up to give the ionic strength as a function of the charge numbers of various combinations of anions and cations as shown below:

Anion	X^-	X^{2-}	X^{3-}	X^{4-}
m^+	1	3	6	10
m^{2+}	3	4	15	12
m^{3+}	6	15	9	42
m^{4+}	10	12	42	16

The Colloidal State

The point emerging from the above table is that although the two salts exist in solution in equal concentration, the ions with higher ionic strength provide more effective screening to the ion or colloid particle. It is possible to define a radius known as the Debye length r_D which determines how strongly the charge on the ion or particle is shielded by the counter-ions. If r_D is large, the shielded potential is similar to the unshielded potential; whereas, if it is small the shielded potential is much smaller than unshielded.

The diagram below shows the solution of the Debye equation for r_D for counter-ions of the type M_nX_m.

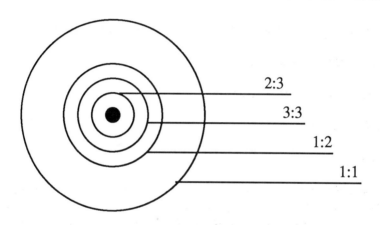

Figure 10

The parameter r_D is important since it defines the "thickness" of the ionic atmosphere.

Typically, for a 1:1 electrolyte in aqueous solution at room temperature with molality = 0.01 mole kg r_D would be 3.0nm.

Applying these ideas to colloids, we find that we can describe the ionic atmosphere around a colloidal particle by the Debye length and that the Debye length decreases as the ionic strength of the medium increases.

At high ionic strength, we have a dense ionic atmosphere around the particle

22 *Fundamental Principles of Sol-Gel Technology*

which causes the potential on the surface of the particle to fall to its average value in the solution after a very short distance.

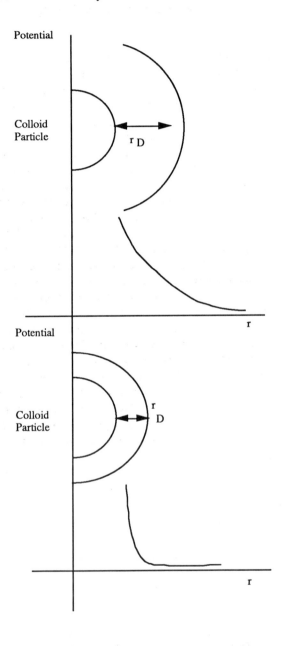

Figure 11

The Colloidal State

If the potential falls rapidly within a short distance, there is very little to stop the close approach of colloidal particles and thus coagulation occurs [precipitate]. In ceramic jargon, we often refer to this as flocculation. This effect is being driven by van der Waal's forces.

Now think of the effect of adding ions to a colloidal suspension. If we add ions of high charge, they will effectively screen the surface charge on the colloid and bring about flocculation even at quite low concentration. The addition of electrolytes has the effect of compressing the diffuse double layer. Not only will they compress the diffuse layer, but they may also absorb close to the particle surface. The increase in electrolyte concentration progressively reduces the double layer interaction and the van der Wal's forces gradually take over causing coagulation. There must be a critical concentration at which the repulsive forces are overcome. This concentration is known as the critical coagulation concentration (CCC). The concept of CCC is not precise but rather an arbitrarily chosen set of conditions. Firstly, we consider the electrolyte to be *inert* which suggests no interaction with the colloidal particle. The concentration at which coagulation just occurs is again arbitrary since one needs to define the *onset of coagulation*; how is it measured? However, provided certain standard conditions are defined, the important point is that the CCC is very dependent on the charge number of the counter-ions. It is almost independent of the ion type also equally independent of the charge number of the co-ions and the concentration of the sol but shows a slight dependence on the nature of the sol. A set of empirical statements have been put together to describe this behaviour known as the Schulze-Hardy rule. Let us examine a practical system, that of a colloidal alumina sol and the progressive addition of a range of potassium salts as electrolytes with the counter-ions increasing from mono to trivalent, i.e., charge number increasing from one to three. In the following table, we can consider the units of concentration as arbitrary. The most significant point is the sharp decrease in the critical coagulation concentration as the counter-ion increases from one to three in charge number.

Alumina Sol Al_2O_3 +ve charges

Salt	CCC
Mono valent anion	
NaCl	43.5
KCl	46

(Continued on p.24)

	KNO$_3$	60
Divalent anion	K$_2$SO$_4$	0.30
	K$_2$Cr$_2$O$_7$	0.63
	K Oxalate	0.69
Trivalent anion	K$_3$Fe(CN)$_6$	0.08

There have been quantitative theories proposed to describe this effect based on energy changes when electrical double layers overlap. I do not propose to discuss these theories in any detail but some of the consequences of these theories will be mentioned where relevant.

It is also possible to model the double layer in more detail by defining substructures as shown below:

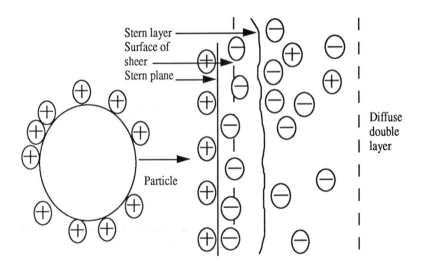

Figure 12

The Colloidal State

It is not relevant here to study the theories of the double layer structure. However, it is interesting to examine diagrams which map the electric potential from the surface of the particle outwards into the solvent. According to the theory stated by Stern, the potential diagram across the particle/solvent profile is as shown below:

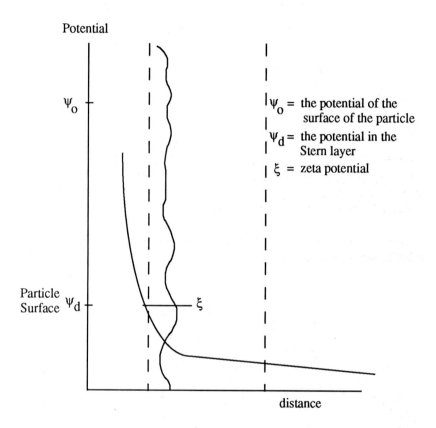

Figure 13

Ions adsorbed are situated within the Stern layer which is the region shown by the boundaries of the surface of the particle and the Stern plane.

This situation depicts a normal double layer in which counter-ion adsorption dominates over co-ion adsorption. It is, however, possible for a charge reversal to take place within the Stern layer, i.e., close to its surface if a polyvalent counter-ion is adsorbed (this also applies to surface active ions but these have not been discussed yet). The potential diagram is now modified as below:

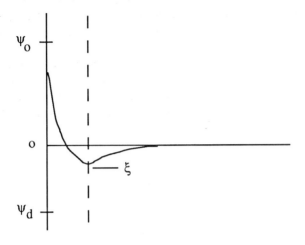

Figure 14

Here ψ_o and ψ_d have opposite signs. Another situation can arise where there is adsorption of surface active co-ions in which case ψ_o and ψ_d will have the same sign but ψ_d is ψ_o as shown below:

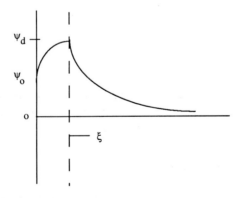

Figure 15

The Colloidal State

We have already discussed the *zeta potential* and *surface of shear*. The zeta potential is the potential between the charged surface and the electrolyte. It is not possible to quantitatively locate the various layers and planes. All we can say is that the surface of shear is located a little further out into the double layer than the Stern plane. Also that the zeta potential is a little less than ξ but it is reasonable to assume these are equal and little is lost in reality by doing this. The only time there is a descrepancy is at high electrolyte concentration when the double layer is compressed. Basically, we are now looking at a very simple interpretation whereby the compression of the diffuse double layer causes a steeper gradient in the potential curve which accentuates the difference between ψ_d and ξ.

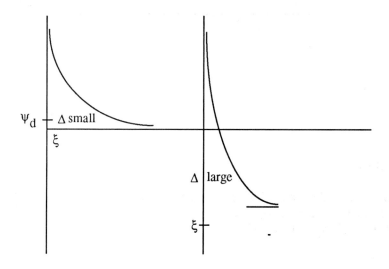

Figure 16

Finally, if the colloid adsorbs a surfactant non-ionic species this would have the effect of displacing the surface of shear quite a long way out from the Stern plane so that the zeta potential is a lot lower than ψ_d.

3.4 COLLOID STABILITY

So far, we have considered the theories and models that describe the double layer but now we need to apply these principles to the stability of colloids. We obviously have competing mechanisms trying to establish a *dynamic* equilibrium. We can generalize this by stating that:

- At small and large interparticle spacing, van der Waal's forces dominate.
- At intermediate separations, double layer repulsive forces dominate.

We can illustrate this by two more potential energy vs. distance curves.

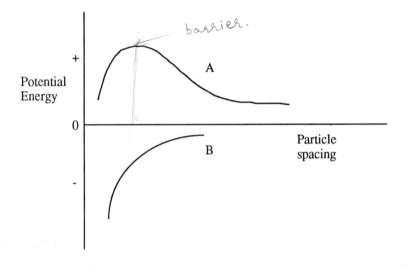

Figure 17

The curve A shows a distance at which the repulsive energy reaches a maximum; whereas, in curve B there is no net repulsion at any particle spacing since the double layer repulsion does not have a strong effect.

Let us now examine curve A. There is a potential barrier here to particles coming close enough to each other to coagulate. What does this barrier depend upon? It depends on:

- ψ_d
- the range of the repulsive forces

If we take a colloidal system and progressively increase the electrolyte concentration, we can expect the following potential energy curves to occur:

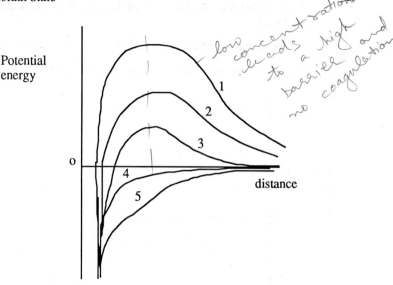

Figure 18

Curves 1 to 5 show the change in potential energy with separation distance for increasing concentration of electrolyte.

If the potential curves are extended to greater distances, there is sometimes a secondary minimum in the curve which can lead to flocculation of particles having large average spacing. One such example of this is iron oxide. If the sol is concentrated but the electrolyte concentration is too low for coagulation in the primary minimum, then we see a separation into two phases and this separation is reversible.

3.4.1 COLLOID STABILITY AND AGITATION

It should now be apparent that colloids are dynamically unstable and hence the ones which are observable are basically slow in their process of coagulation. We have seen how the double layer significantly influences the desire for two particles to come close enough for van der Waal's forces to cause them to coalesce and how the structure of this double layer is affected by addition of various ions into the solution.

Throughout the life of the colloid its worst enemy, from the point of view of stability, is thermal motion. The greater the thermal activity in solution, the greater the probability of close approach of colloid particles. This is easily verified by heating sols which leads to flocculation. We again have an expression for the stability of a colloid under pure thermal agitation; it is under *perikinetic* conditions.

However, if we wish to encourage aggregation of the colloid particles, we can influence this by agitating the molecules in the sol externally. Particularly with large particles, agitation can increase the rate of aggregation by up to 10,000 times. Agitated conditions are known as *orthokinetic*. Use is made of this fact in some industrial processes.

3.4.2 PEPTIZATION

As you plough through the literature connected with sol-gel technology, you will encounter this expression time and time again and the phenomenon is very important so we will deal with it in some detail here.

Peptization is a method of achieving a good dispersion in solution by controlling the composition of the dispersing solution. It is not an isolated process but is often associated with coagulation and is also a time dependent process. This is achieved by changing the composition of the dispersing medium. Often this is done by adding polyvalent co-ions to the solvent. It is also possible to add a surfactant or to simply dilute the dispersing medium. The electric double layer is modified such that a potential barrier is created against recoagulation. The sedimentation volume of a peptized sol is small compared to aggregated particles.

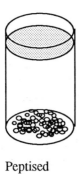
Peptised
sediment

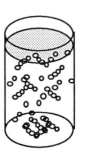
aggregated
particles

Figure 19

The Colloidal State

In colloidal oxide sols such as alumina (boehmite), nitric acid is frequently used as the peptizing agent.

3.5 EPILOGUE

So far, we have considered the formation of colloidal sols from the point of view of the colloid or physical chemist. There are certain conceptual ideas that should emerge from this chapter which should give insight into the effects of:

- warming up colloidal suspensions
- adding electrolytes to colloidal sols
- agitating colloidal sols
- understanding a little about the starting materials

Having formed the sol we have been preoccupied, up to this stage, in stabilizing the sol but so far we have merely ended up with some form of inorganic material in solution. In this state, it is a useful starting point but we now have to get it from a solution into a useable form:

- a coating
- a powder
- fibres
- monolithic shapes

With regard to the formation of powders, we can envisage many ways of destabilizing the colloid in order to grow particles large enough to precipitate out and form powders. The state of aggregation of the powders and the morphology and size distribution is an area that can be controlled but this still only yields a precursor material. The great excitement that has arisen in the recent years of sol-gel development is the possibility of using the process to achieve "the finished product" which can include:

- coatings on substrate of all types including plastics (an example here is a scratch resistant coating of inorganic material on a plastic lens)
- fibres - there is great activity in the field of optical fibres since one can chemically formulate the glass with the desired refractive index to produce optical fibres of high quality and low db transmission loss
- near net shapes - with careful selection of the starting system and a

great deal of Alchemy it is possible to produce large monolithic pieces of glass and ceramic

In the following brief chapter, we will concentrate on the way we can achieve a system from the sol that can be turned into the finished product. The route is through the conversion of the Sol to the Gel!

4 SOL-GEL TRANSITION

4.1 INTRODUCTION

Having achieved a complex twenty-five component sol, stable for 1,000 years, which defies all phase diagrams, we need to convert this animal from *liquid* to *solid*. Here begins the next series of problems. Again we will consider what is known and generally accepted by physical chemists on this subject, in the hope that it gives some enlightenment, but when we move on to proven practical systems we will have to support the pragmatic with the didactic rather than use the didactic as the predictive driving force.

I suppose I am saying that later in this monograph you will encounter many recipes which may not fit into one of the more simplified chemical processes I have tried to describe. I hope you will not totally reject the theory but try to use theory to generate a qualitative feel for "what works and what doesn't".

So now we move on to the next stage which is very interesting and that is the conversion of the Sol to a Gel.

4.2 THE FORMATION OF GELS

Having considered the aggregation of particles in a sol as if they were weightless suspended bodies, we should recognize that in most inorganic systems the solvent will have a density of between 1 and 1.5 whilst the solid will be between 2.2 and 3.8. If we then allow enough aggregation to take place, the solid particles will want to settle out of the solution. Depending on the microstructure of the precipitate, the solid coming out of solution will occupy a certain volume in solution and this final volume must depend upon the state of aggregation of the colloidal particles in solution

Let us now progress through three extremes:

1. Large, peptized particles pack densely to form a sediment which will not re-disperse.
2. Particles which are partially aggregated have bridging capability and therefore give a loose sediment which can often be re-dispersed.
3. In many cases the volume of this sediment may well equal the volume of the complete solution.

In this third case, the macromolecules or particles have aggregated in such a way as to form a continuous three dimensional network structure throughout the whole volume of the original solution.

The schematic diagram below tries to represent each case:

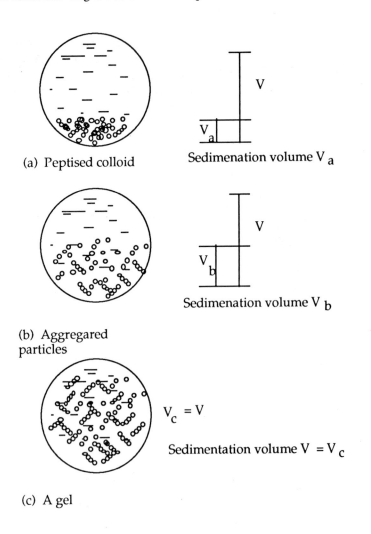

Figure 20

In the third case where the sedimentation volume is equal to the volume of the

Sol-Gel Transition

original solution, we have a state in which the particles or macromolecules have cross-linked and formed a structure which is capable of immobilising the remaining solvent and this gives us a solid structure which often exhibits the visco-elastic properties of a gelatinous mass. This state is called a GEL. This is the stage that most interests your children!

The mechanical state of this gel depends very much upon the number of cross-links in the network. In many ways, we can interpret "cross-linking" by considering simple polymers. Polymeric molecules tend to be long chain molecules of high molecular weight. In the solid, these molecules are basically mixed up and held together by van der Waals forces.

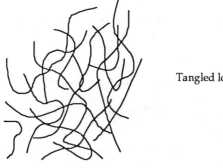

Tangled long chain molecules

Figure 21

Some plastics still have functional groups attached to the long chain, e.g.,

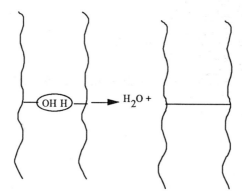

Figure 22

With a little help from thermal energy, and a catalyst, it is possible to get functional groups between adjacent molecules to react and form a chemical bond between the molecules. This intermolecular bond is then referred to as a "cross-link" and polymers that exhibit this behaviour are referred to as *thermosetting* plastics.

It is obvious that the greater the degree of cross-linking, the more rigid the structure will become. If we consider a gel composed of cross-linked long chain molecules rather than colloidal particles with cross links, we find the physical chemistry of these systems is similar but not quite the same since often a slight change in conditions in the sol, e.g. pH, can transform a solution of macromolecules into an aggregated colloid.

You will see this situation arising when we examine the effect of acid and base catalysts on organometallic systems. However, much of the chemistry we have considered to date is still very relevant but it requires application to more complex systems than the models we have examined up to now. In the previous chapter, we concentrated very much on the stabilizing effect of the double layer on the sol. If we consider a structure comprising macromolecules, we often need to also consider solvation effects. Many macromolecular solutions contain lyophilic (solvent loving) molecules. A good example is that of gelatin. It has such a strong affinity for water that it forms a stable gel under almost any conditions including children's birthday parties.

You will remember earlier that we discussed the isoelectric point as a pH level at which there is basically no double layer effect between particles in solution. Because of this, we would expect many systems to flocculate or precipitate out but gelatin has such a strong attraction to water molecules that it remains in a stable gel form. There are many other systems that exhibit this behaviour. In the previous chapter, we also saw how the addition of small amounts of electrolyte had pronounced effects upon the charge on the colloidal particle and the structure of the double layer. In the case of these highly solvated macromolecular solutions, we can often load the sol with large amounts of electrolyte before any coagulation and precipitation occurs.

At high concentration, the ions of the electrolyte compete with the macromolecules for the solvent molecules and effectively dehydrate the macromolecules. The tendency to do this depends upon the ions' desire to become hydrated which in turn is effected by the ion size and charge number. It is possble to list some cations and anions in decreasing order of efficiency with respect to causing precipitation in macromolecular solutions.

Sol-Gel Transition

Cations: $Mg^{2+} > Ca^{2+} > Sr^{2+} > Ba^{2+} > Li^+ > Na^+ > K^+ > NH_4^+ > Rb^+ > Cs^+$

Anions: $Citrate^{3-} > SO_4^{2-} > Cl^- > NO_3^- > I^- > CNS^-$

We will find that the double layer effect does not give a complete explanation of the behaviour of macromolecular solutions and thus further stabilization mechanisms have to be considered. As we move on to these complex systems we find that the theoretical models become less quantitative.

One general explanation of the stabilization of macromolecular solutions is that of *steric stabilization*. This term can cover a multitude of mechanisms. Generally, the macromolecules have both lyophobic and lyophilic parts. If the solvent molecules have a strong affinity for the lyophilic functional groups on the polymeric molecules, this creates aggregation of the molecules and likewise if there is little attraction to the solvent the opposite occurs. We will see later how this effect is utilized in causing a gel to precipitate out by the addition of species which decrease the tendency for the macromolecules to solvate.

We can delve a little deeper into the thermodynamic arguments here! Let us consider free energy changes taking place in the macromolecular solution. The free energy must be a function of:

- temperature
- pressure
- solvent type

We need to refer back to the concepts of ideal and non-ideal solutions. Macromolecular solutions tend to be non-ideal because they interact with the solvent and there is a large excluded volume in the solution. The contributions to non-ideality are thus two fold:

- solvent interaction effects
- excluded volume effects

The solvent interactive effects cause changes in enthalpy and are thus called enthalpic; whereas, excluded volume effects alter the entropy (entropic).

There is a point at which the free energy change in the solution is zero and at which the solution behaves as an ideal solution. At this point, the enthalpic effects cancel the entropic effects. We call this point the θ (theta) point and the temperature at which it occurs the theta temperature.

If we write down the simple free energy equation:

$$\Delta G = \Delta H - T\Delta S$$

ΔG = free energy change

ΔH = enthalpy change

ΔS = entropy change

T = temperature

then the steric stabilization $+\Delta G$ could be a result of a positive enthalpy change or a negative entropy change or both. How can we interpret these changes?

- $+\Delta H$ means we have a release of energy as bound solvent molecules are forced out of the polymeric chains when they interpenetrate.

- $-\Delta S$ means that the degree of freedom for the molecule to change its configuration is reduced as the chains interpenetrate and prevent this happening.

This now gives rise to three distinct conditions:

a) If ΔH is +ve, the dispersion will be sterically stable over the whole temperature range.

b) If ΔH +ve and ΔS +ve, the dispersion will flocculate on heating above the Œ temperature (enthalpic stabilization).

c) If ΔH -ve and ΔS -ve, the dispersion will flocculate on cooling below the Œ temperature (entropic stabilization).

If we look at actual systems, we find that the majority of macromolecular sols in aqueous solvents follow enthalpic stabilization; whereas, those in non-aqueous solvents are entropically stabilized.

At this point, we have covered most of the relevant physical chemistry describing the behaviour of colloids and macromolecules in solution. I now intend to move on to discuss actual systems.

Firstly, we will examine true colloidal sols and see how these are achieved and how the sol can be used to prepare a powder or be transformed into a gel. We will then examine the more common cases of gel formation via organometallic compounds such as alkoxides. Whilst the physical chemistry will give an appreciation of why certain behaviour is taking place, you will find that many "real" systems are complex and thus the formation of sols and gels tend to appear as a recipe often individual to that particular system.

5 FORMATION OF SOLS

Throughout this chapter, I will make reference work to the research of many scientists. Where you find particular systems of interest, I recommend that you obtain the relevant publication and study this in depth. Again in this chapter, it will be useful to distinguish between the true colloidal particulate sols and the systems which result from interpenetration and cross-linking of macromolecules.

There are basically two routes to the formation of colloidal sols. One consists of breaking a material down from a macroscopic to a microscopic state and this can be achieved by a variety of methods:

- milling or grinding the material down
- breakdown of larger aggregates by ultrasonic dispersion
- discharge between two electrodes made of the material in a solvent
- electrolytic deposition

Of the above methods, milling is perhaps the most commonly used but there are inherent difficulties in using this technique. There are many forms of mill.

5.1 BALL OR JAR MILLS

These consist of ceramic, metal or rubber jars part filled with ceramic balls in which we put the material to be milled. The jar is rotated on rollers so that the ball charge tumbles around and breaks up the particles.

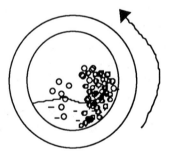

Figure 23

Formation of Sols

As with any technique, there is both art and science associated with milling. For instance, the efficiency of the ball mill depends upon many factors:

1. The ratio of material to be ground to grinding media.
2. The hardness of the media.
3. The size distribution of balls in the media.
4. The rotational speed of the mill.
5. The use of liquids to suspend and disperse the particles and prevent aggregation.
6. The use of electrolytes.
7. The change in electrolyte concentration as an ion is specifically leached from the surface of the particles being milled.
8. Scaling up effects.

In industry, because of the complexity of such a process, there is mainly a high level of operator skill backed up by modicom of science. Each operator, whilst striving for efficiency, will always come up against the law of diminishing returns. Milling reaches a point where efficiency asymptotes off to a level where prolonged milling has little effect in reducing the particle size any further. This is because the finer the particles, the more they tend to aggregate. To some extent, this can be manipulated by the use of liquids to disperse the particles but liquids only have the effect of moving the onset of the asymptote shorter up the time scale. Liquids can have two effects:

- Inert liquids help separate particles.
- Surfactant liquids can be introduced for wet milling to again alter the double layer and separate the particles.

Agglomeration still eventually occurs. The figure on p.42 schematically represents the agglomeration mechanism for alumina particles being wet milled in water: The particle in (a) is water solvated with the hydrogen ions preferentially absorbed on the surface in a double layer effect. Figure (b) depicts the type of mechanism taking place on a fresh fracture surface exposed by milling. H^+ ions are exchanged for Al^{3+} ions on the surface. This forms a monohydrate layer and Al^{3+} ions dissolve in the water.

As the reaction proceeds, the Al^{3+} ions form a gel in the solution. This gel becomes a binding phase. Other materials exhibit their own specific mechanisms. More sophisticated milling often gives a similar minimum particle size but may achieve this in a much shorter time. These milling techniques also give

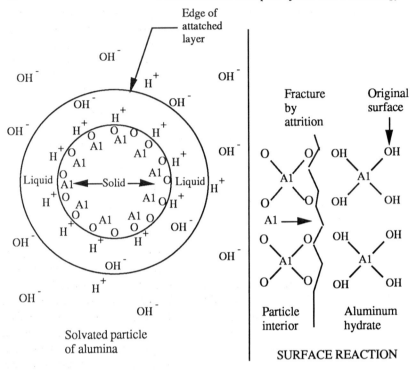

Figure 24

different size distribution to ball milling. These techniques include:

- vibro energy milling: here the jar with charge and media is vibrated rather than rolled

- fluid energy milling: tangential air jets are used to create a fast moving annular air flow in a chamber. The coarse powder is placed in the vortex and mills by collision with itself

- ultrasonic dispersion: use of short wavelength ultrasound to cause cavitation between particles and separate them by agitation

Formation of Sols

- discharge at electrodes: we prepare electrodes from the material we wish to disperse and create a discharge between them, either in air or in an inert non-flammable fluid, this has the effect of eroding small particles from the electrodes

Attrition techniques are not the best routes to producing colloids. They tend to be lengthy and produce a wide range of particle sizes.

Rather than breaking coarse particles down, another series of techniques involve building the particles up from smaller ones. There are many methods and I will briefly describe them here:

1 *Flame Oxidation*: This technique involves the oxidation of a salt in a high temperature flame. The process has been used for many years to produce colloidal silica. Under the commercial name of Cab-o-sil a very fine silica "soot" is formed by the flame oxidation of $SiCl_4$. Other materials such as Al_2O_3 and TiO_2 have also been prepared this way. This technique, however, still yields a broad particle size distribution. There are many modifications to this technique, for instance, oxidation and hydroysis can take place in a flame. We do not necessarily need to use a flame, a plasma or heated tube can be used instead. Not only can we oxidize or hydrolyse halides; other chemicals such as organometallics can be used as precursors. These techniques generally produce particles in the size range 20-100nm and with surface area between 50-400m^2/g.

2 *Chemical Routes*: In recent years, attention has been paid to chemical routes for producing sols of which there are now many. These routes involve the development of an insoluble species by a chemical reaction in solution. The form of the solid particle is a function of:

- the nature of the reaction
- the concentration of reactants
- the reaction conditions of temperature, pH, etc.

Because of the wide range of variables, this gives quite a degree of control so that the versatility of these chemical methods gives considerable scope for the preparation of useful precursor materials. An important process in this category is that of hydrolysis of organometallics, in particular alkoxides. This method is commonly referred to as sol-gel technique and because of its importance it will be dealt with in much greater detail later. In particular, we will examine how amorphous gels can be prepared from sols synthesized by this route. Firstly, I will list the range of chemical techniques briefly.

Precipitation techniques

Provided that the chemical one is trying to suspend as a sol has sparing solubility, it is often possible to react a soluble form such as a salt in a solution in order to develop the insoluble precipitate.

Generally, the reaction is one of hydrolysis which generates the hydroxide or hydrated oxide. Quite often hydrous oxides will precipitate out in the form of a gel which can then be used in this state or further processed to coagulate the particles from the gel to form a precipitate.

Other precipitation techniques involve the use of citrates, oxalates and other salts or alkoxides. A great advantage of forming colloids and precipitates from solutions is that the particle distribution can be made very narrow. It is important to be able to deaggregate any precipitate which forms if the sol is to be stable and further processed into a granular form which is often desirable where ceramic precursor powders are needed. Such process have been developed to sophisticated levels by Harwell (8). One example is the precipitation of hydrous cerium oxide which is performed by adding $NH_4OH - H_2O_2$ to a suitable cerium salt. After washing the precipitate to remove the salt, we find that the precipitate consists of small particles 0.1 µm which are aggregated into much larger particles.

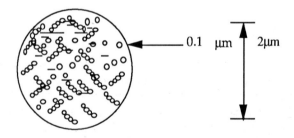

Figure 25

In fact, the 0.1 µm blobs are also built up from smaller 8nm particles. The deaggregation is brought about by protonation of the particles using nitric acid. The ratio of protons to Ce(IV) ions is very important. If we add too much acid,

Formation of Sols

we can dissolve the cerium hydroxide and if we add too little, the bridging hydroxyls between particles is not broken. Hence, just the right amount of acid must be added. After treatment, the cerium hydroxide still remains insoluble. This is because the anion associated with the H^+, i.e., NO^{3+} now helps to flocculate the deaggregated particles. These anions bring about a decrease in thickness of the electrical double layer. Simply by removing the anions in the excess liquid the resulting flocculated cerium hydrous oxide is easily dispersed in water to form a transparent sol. Other systems can be prepared in a similar way, for instance, a good starting point for a chromium hyrdroxide sol is chrome alum $KCr(SO_4)_2$ which can be aged to form the sol at 75°C in 24 hours. Ageing simply means leaving the solution for a period of time to allow something to happen.

Particle morphology can also be controlled. For instance, if we consider the preparation of colloidal hydrous alumina our starting material can determine the shape of the colloidal particles in the sol. Aluminium sulphate can be used to prepare spherical hydrous alumina particles; whereas, chlorides, nitrates and perchlorates yield anisotropic sols consisting of ellipsoids, rods or polyhedra.

Monodispersed sols of iron and copper are also readily obtained by ageing various salts such as ferric nitrate and sodium sulphate or in the case of copper a modified Fehling's solution.

Silica can be formed by the addition of sulphuric acid to sodium silicate. Here the acid has the role of removing water. Either precipitated silicas can be formed or a modification of this technique can be used to prepare monolithic silica pieces.

Solvent extraction

We can illustrate this process by another example. If we are preparing a titanium dioxide sol we can begin with a concentrated aqueous acidic solution of titanium chloride. This is an ionic solution. Now the precipitation of crystalline titania from the solution is found to occur at a particular Cl:Ti ratio, Cl:Ti~1.5. Thus, it is necessary to remove the hydrolytic acid from the solution. This can be achieved with an immiscible organic amine. $R_3C - NH_2$. This amine can combine with the anions of the acid and form nitrates or chlorides depending on the acid present. This enables fairly concentrated sols to be prepared.

Hydrolysis of organometallics

Often the hydrolysis of alkoxide solutions at pH < 7, if carried out under controlled conditions, leads to a gel which is a complex interlinked polymeric structure. It is, however, possible to use alkoxides to prepare true colloidal sols.

Some early work involved the production of silica sols consisting of monodispersed spheres. This process was named after one of the inventors and is now commonly referred to as the Stoeber process.

Taking an alkoxide such as tetraethoxy orthosilicate (you will see many forms of naming alkoxides but we will try to be consistent), the solution can be hydrolysed by bases as well as of acids. Ammonia is a common base for this purpose.

Stoeber discovered many things about the growth of the particles including a relationship between size of particles and the amounts of water and ammonia present as shown in the figure below:

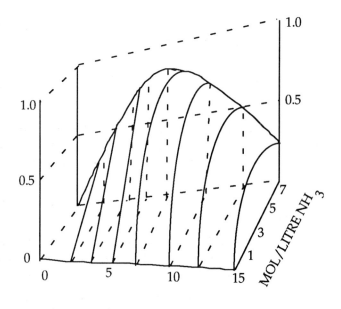

MOL/LITRE H_2O

Figure 26

Formation of Sols

He also discovered the fact that the alkyl group determined the reaction rates; methyl being the fastest and tetrapentyl the slowest. These reaction rate changes were paralleled in the solvents. For instance, the reaction took place much faster in methanol than n-butanol.

Anion extraction with epoxy compounds

Another more recent development reported by UKAEA, Harwell (9) involves the preparation of "dispersible gels" leading to the formation of sols. This is achieved by the use of epoxy compounds which undergo the typical reaction:

$$MX + CH_2 - CH_2 + H_2O \longrightarrow MOH + HOCH_2CH_2X$$
$$\diagdown \quad \diagup$$
$$O$$

There are many practical examples of this technique which are described in detail in their patent. A few are outlined below:

1. A gel was formed by treating a basic zirconium chloride solution with propylene oxide. This gel was dispersed into a sol by shaking in water.

2. Aluminium chlorohydrate solution in methylated spirit was treated with propylene oxide and gelation occurred. At the correct propylene oxide concentration, the gels were readily converted to sols by redispersing in water.

3. Aqueous titanium tetrachloride in methanol was treated with glycol to produce a water dispersible gel.

4. Nickel chloride hexahydrate in methanol was treated with propylene oxide on which a gel was formed. This gel was dispersed to a sol in ethylene glycol monoethyl ether. (Trade name is Cellosolve.)

Rather than add to the list of sols from solutions at this stage we will deal with specific examples when we examine sols and gels from alkoxides in greater detail. Finally another route to colloids is possible.

3. *Thermal Decomposition of Metal Salts*: Let us examine what happens when we heat a hydrolysable metal salt. In the example here we consider heating hydrated thorium nitrate, $Th(NO_3)_4 6H_2O$. Firstly, the salt melts in its own water of crystallization at 110°C and nitric acid is evolved according to the following reaction:

$$Th(NO_3)_4 + XH_2O \longrightarrow Th(OH)_{4-x} + XHNO_3$$

When the NO_3:Th ratio reaches 1 the temperature has increased to 180°C. The increase in temperature is due to the polymerisation of the hydrolysed Th(IV) ions. If heating progresses to 480°C, further nitrate is lost. At this temperature the NO_3:Th ratio is 0.1 and this material will readily disperse in water to form a concentrated sol with crystalline particles of between 10-15nm. A similar dispersible oxide can be produced from the oxalate.

Having seen how we can form stable sols and discussed the importance of conversion of sols to gels, we will now deal with this transition in some detail since it represents a considerable effort nowadays in the preparation of ceramic materials and glasses.

6 SOL-GEL TRANSITION IN ALKOXIDES

In this section, I will concentrate very much on the gels that arise from the hydrolysis and condensation reaction of alkoxides. In the early days of sol-gel activity, silica was chosen as the model system being a well known material in its naturally occurring crystalline and amorphous states. I suggest here that we do not depart from tradition and in attempting to understand the sol-gel transition, we consider silica as a typical system. From now on, we will use the expression alkoxide frequently and it is essential that you understand the expression at this stage.

In general terms a metal alkoxide has the chemical formula:

$$M(OR)_n$$

where M = metal

R = alkyl group   ie CH_3.methyl C_2H_5.ethyl
C_3H_7.propyl C_4H_9.butyl

n = the valence of the metal atom

One could argue that silicon was not a metal but it does exhibit metallic behaviour and like aluminium has been considered for many years to be on that boundary between metallic and non-metallic elements. This is particularly noticeable if we look at the oxides of the Group IV elements. They are known as "amphoteric oxides" which means they exhibit both acidic and basic characteristics. For our purposes, we will consider silicon as a tetravalent cation, Si^{4+}. A very common alkoxide of silicon is TEOS or tetra-ethyl-orthosilicate with the formula:

$Si(OC_2H_5)_4$

Let us now introduce some water into the TEOS. Under certain conditions hydrolysis of the orthosilicate will occur:

$Si(OC_2H_5)_4 + H_2O \longrightarrow Si(OH)(OC_2H_5)_3 + C_2H_5OH$

or

$Si(OC_2H_5)_4 + 2H_2O \longrightarrow Si(OH)_2(OC_2H_5)_2 + 2C_2H_5OH$

or

$Si(OC_2H_5)_4 + 3H_2O \longrightarrow Si(OH)_3(OC_2H_5) + 3C_2H_5OH$

and finally

$$Si(OC_2H_5)_4 + 4H_2O \longrightarrow Si(OH)_4 + 4C_2H_5OH$$

This hydrolysis reaction is catalysed by the addition of an acid or base. In fact, the final form of the hydrolysed silica is very dependent on the pH of the solution. The conversion from the intermediate state to metal oxide is by loss of alcohol and water. So in the case of silicon:

$$Si(OH)_3(OC_2H_5) \quad SiO_2 + 2H_2O + C_2H_5OH$$

Let us examine the effect of pH and water content on the gel formation from a silica sol. At low pH levels, i.e., very acidic, the silica tends to form linear molecules which are occasionally cross-linked as shown in the figure below:

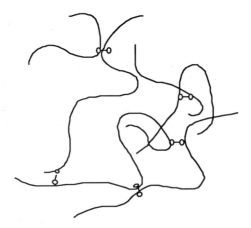

Figure 27

This fact is readily demonstrated since diethyl ether dissolves in strong mineral acids. Protons which seek regions of high electron density are "electrophilic". Referring back to silica, we can see that the oxygen atom on the alkoxide group will attract electrophilic reagents or protons H^+ in this case.

Here the linear chains have a low density of crosslinks. This leads to a soft gel which is reversible in that it can be redispersed in solution. As we decrease the pH value we increase the number of cross links between polymer chains. As we decrease the pH we find that the linear polymers become more branched and that

Sol-Gel Transition in Alkoxides

the number of cross links increases. The differences between lower and higher pH here tend to be not so sensitive to the water content just the pH value. When we move onto base catalysed systems we begin to see the importance of water content. Staying at the low pH end (pH 1 - 3) we can see a transition occurring. At low pH (highly acidic) hydrolysis occurs by electrophilic attack on the alkoxide group. To understand this term we need to go back to some basic organic chemistry.

Whilst we generally consider most organic molecules to be electrically neutral there is often an electrical polarization in the different parts of the molecule. If a molecule contains an atom with a partial negative charge (high electron density) then this particular atom is very vulnerable to attack by an atom or ion that has an attraction for electrons, these groups or atoms are known as *electrophilic*. Take the case of diethyl ether. Here the oxygen atom has a high electron density owing to its strongly electronegative character and the polarizing effects of the alkyl groups. The oxygen atom also possesses two pairs of unshared electrons, so the oxygen atom becomes a very favourable site for undergoing co-ordination with protons (H^+) from the acid.

There is obviously a converse situation where the oxygen attracting high electron density leaves some connected atom partially electron defficient and thus positively charged. These groups or atoms are prone to attack by lovers of positive charge and these reagents are known as *nucleophiles*. They behave as electron donors by either;

- transferring electrons to the external atom or ion, or

- by sharing their electrons with an atom or ion

Molecules or ions containing an atom with an unshared pair of electrons commonly act as nucleophiles. We can list common electro and nucleophilic reagents as follows:

Typical Electrophilic Reagents

H^+, H_3O^+, Halogens, Halogen acids, e.g., HCl, HBr, HOCl, HNO_3, H_2SO_4

Typical Nucleophilic Reagents

OH- and many other negative ions, H_2O, alcohols, NH_3, NH_2OH, $C_6H_5NHNH_2$

In general terms we can regard all electrophilic reagents as oxidizing agents and nucleophilic reagents as reducing agents.

Going back to the case of silicon, we find that at higher pH (increased basicity) hydrolysis and polymerization occur by nucleophilic attack on the silicon ion (Si^{4+}). In either case, the result is a hydrated gel which is not a true colloidal dispersion of silica but rather a particular form of three dimensional polymer.

It is only when we go to very high pH and excess water content that a true colloidal dispersion forms. The explanation for this behaviour is not too complicated if considered stepwise:

High pH in alkoxide with excess of water.

Extensive hydrolysis and polmerization.

The linear chain molecules (siloxanes) under these conditions become soluble.

The chains formed rapidly depolymerize (solubility is related to molecular weight so lower MW segments of polymer chains are more soluble than higher MW segments).

Small segments dissolve and redeposit onto the larger chains so the smaller molecules decrease in number but assist the large molecules to grow. This process is called Ostwald ripening.

You will come across this behaviour in many other areas. What happens is that the number of growing particles decreases but those that grow do so to a sensible size. This situation is very similar when one considers the nucleation of crystal growth in glass-ceramics. The ripening effect leads to the formation of colloidal silica particles.

It is not difficult to understand the way that low molecular weight species sacrifice themselves in order to grow on larger particles if we simply refer back to surface free energy. Any small species will quickly want to be swallowed up by the larger one in order to reduce surface free energy. Perhaps one of the most striking examples of this driving force is the example of two soap bubbles connected by a tube but where the bubbles are of unequal size. If one asks the general question "Will the larger bubble get smaller and the two eventually reach equilibrium in size?", many people would intuitively say "yes". But the reverse is true. The smaller bubble gets smaller and the larger one larger. It is not important that a bubble consists of a skin of liquid around gas or that a particle consists of an accumulation of atoms in a liquid. The common fact is that each entity tries to reduce surface free energy and, therefore, surface area. In this chapter, we are more interested in gels than sols or colloidal precipitates since

Sol-Gel Transition in Alkoxides

we can use gels as precursors for:

- forming glasses which would otherwise be almost impossible to achieve by fusion
- coatings onto substrates of all types
- monolithic precursors - optical fibres drawn from duplex rod prepared by sol-gel
- fibres drawn from the gel when it reaches the optimum rheological characteristics
- near net shapes - using techniques such as drying control additives to achieve monoliths that require a minimum of post forming machining

So we consider the gel state as most useful here, whereas, the ceramist requiring powders of specific:

- homogeneity
- particle size and distribution
- particle shape

is only interested in gels as an intermediate to the above precursors. The "gel" state can, however, solve many problems and is therefore studied in detail here. It is most useful to obtain the gel in an amorphous state and technique can ensure this in most cases. We have seen how a pure silica gel can be obtained and whilst silica is a useful material both for research purposes in studying gelation behaviour and in many applications, the most useful systems are often multicomponent oxides. The techniques for their preparation are varied and often require some skill to produce homogeneous gels as well as a scientific understanding of the chemistry. This is due mainly to two factors:

Miscibility and Hydrolysis rate

Not all components are miscible in all quantities. A simple example is the miscibility of TEOS water and alcohol.

In the figure on p.54, we see the original region of miscibility shown as A. We can, however, increase this miscibility region by adding acid. This now increases the miscibility region to A + B

Another factor which can present a problem in a multicomponent system is if each species hydrolyses at a different rate. If one component hydrolyses rapidly

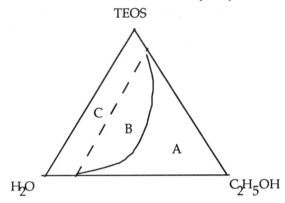

Figure 28

compared to another, this can lead to precipitation of this component in the gel. This problem can be overcome by very slow addition of water or by techniques such as chelation of the cation which will be discussed later.

Let us now examine the synthesis routes for multicomponent gels.

6.1 ALKOXIDE SYNTHESIS

The simplest approach here is to make a mixture of the alkoxides of all the components often in an alcohol with the same alkyl group as the alkoxides. This solution is then reacted with water. Most alkoxides hydrolyse readily but there are a few exceptions. Silicon alkoxide hydrolyses very slowly and requires acid or base catalysts. Trialklyl phosphates are also difficult to hydrolyse.

The hydrolysis reactions are very complicated in multicomponent systems. Firstly, the metal alkoxide groups hydrolyse to form hydroxides. These groups then condense with each other or unreacted alkoxide groups which then form M_1-O-M_2 links (M = metal ion). This produces complex polymers. As the polymer molecules increase in molecular weight and degree of cross-linking, the polymer eventually becomes insoluble and forms a gel.

Silicon is the basis of the most common range of glasses, the silicates, yet it is not the most useful akoxide because of its slow hydrolysis rate. Multicomponent silicates are often prepared by very slow hydrolysis which prevents the more readily hydrolysable alkoxides from precipitating as oxides. One method of doing this is to expose the solution to moisture in the atmosphere. There are also "tricks of the trade" such as placing Cling Film over the beaker and controlling

Sol-Gel Transition in Alkoxides

the rate of moisture adsorption by piercing pinholes in the film. Another method is to partially react with silicon alkoxide prior to mixing in the other alkoxides. Here the reaction can be represented by:

$$(RO)_4Si + H_2O \longrightarrow (RO)_3SiOH + ROH$$

If another alkoxide is now added and this reacts with the silanol forming a soluble metallosiloxane:

$$(RO)_3SiOH + M(OR)_x \longrightarrow (RO)_3Si\text{-}O\text{-}M(OR)_{x-1} + ROH$$

The majority of metal alkoxides will react in this way with partially hydrolysed silica provided there are enough silanol groups present. Generally in silicate glasses, silica is present in proportions greater than the other components so it is not normally a problem.

So far we have referred in general terms to alkoxides but the alkyl group has a strong influence on rate of hydrolysis which is most rapid with methyl and reduces as we go to higher alkyls. This rate is further reduced by the use of branched alkoxy groups.

The table below shows the different reaction rates of hydrolysis as a function of chain length and degree of branching of the alkyl group. This effect on hydrolysis rate is a steric or crowding effect. We can also have mixed alkoxides $(RO)_x(R'O)_{4-x}Si$. Here we find that if R' is the higher alkoxide it will only slow down hydrolysis when $X = 0$ or 1; whereas, when R is a branched group it can slow down hyrolysis when $X = 2$.

Reaction rates for tetra alkoxysilanes $(RO)_4Si$ at 20°C:

R	Reaction rate k
C_2H_5	5.1
C_4H_9	1.9
C_6H_{13}	0.8
$(CH_3)_2CH(CH_2)_3CH(CH_3)CH_2$	0.3

Another effect, which is a function of the nature of the alkyl group, is the inductive effect. Consider an alkoxide with mixed alkyl groups, e.g. methyl ethoxysilane $(CH_3)x(C_2H_5O)_{4-x}Si$ where x is O to 3. If we continue substituting alkyl for alkoxy groups, this increases the electron providing effect of the ligand, whereas, hydrolysis, i.e. substituting OH for OR, increases stability of positively charged species and conversely with hydroxyl substitution. This means that under acid conditions the hydrolysis rate increases with increasing degree of alkyl substitution but under basic conditions the hydrolysis rate increases with hydroxyl substitution. In general terms, as we increase the hydrolysis substitution steps under acidic conditions there will be a tendency for the rate of further hydrolysis steps to decrease, whereas, under basic conditions the increased electron withdrawing tendency of OH and OSi compared to OR can increase the rate of subsequent hydrolysis steps. This effect is inductive. If you are not familiar with inductive effects, let us look at a simple example. In a covalent bond between unlike atoms, the electron pair in the bond is not shared equally between the atoms. The pair tend to be attracted more towards the electronegative atom. Typically, in an alkyl halide the electron density is higher near the chlorine atom than the carbon atom:

$$\diagdown\!\!\!\!\!\!\!\underset{\diagup}{C}\longrightarrow Cl \qquad\qquad \diagdown\!\!\!\!\!\!\!\underset{\diagup}{\overset{\delta+}{C}}\text{———}\overset{\delta-}{Cl}$$

In the event of a series of carbon atoms in the chain:

$$C_3 \longrightarrow C_2 \longrightarrow C_1 \longrightarrow Cl$$

he effect is passed on such that electrons are induced towards the chlorine atoms. Most atoms or groups attached to a carbon atom exert such an effect "the inductive effect" and are electron withdrawing but alkyl groups are electron donating.

Finally, solvents can have a significant effect on rate of hydrolysis depending upon whether the solvent is polar or non-polar or again if we are considering acid or base catalysis. An interesting solvent other than the alcohol is the addition of formamide ($H.CO.NH_2$) to an alkoxide solution. Formamide has large permanent dipoles and therefore is more likely to hydrogen bond more strongly than the alcohol solvent to protons under acidic conditions and hydroxyls under basic conditions. Also the viscosity of formamide is much higher than most alcohols so these combined factors add up to formamide reducing the hydrolysis rate.

Sol-Gel Transition in Alkoxides

So far we have applied various arguments to hydrolysis but similar effects are noticed in the ongoing condensation reations.

Condensation is the polymerisation reaction that forms siloxane bonds due to either RO or H_2O reacting with the hydrolysed alkoxide. In aqueous systems, the sequence is probably a progressive build up of chains from the monomer through dimer, trimer, cyclic trimer, tetramer, cyclic tetramer, etc. and these polymeric rings form the nucleus for colloidal particle formation. The rings are formed by a dynamic process where a reverse reaction, depolymerization and ring opening takes place. Provided there is an abundance of monomers they will attach on and increase the size of the rings. In alcoholic solutions at low pH the depolymerization rate is, however, very low so once a siloxane bond is formed it cannot be easily broken. Here the condensation resembles the normal reaction one would associate with a thermosetting plastic resulting in a three-dimensional network structure which leads to the characteristic gel.

6.2 ALKOXIDE SALT METHODS

The formation of multicomponent gels can be achieved without the use of all alkoxides and for reasons of economics it is often desirable to reduce the number of alkoxides in the system. In many cases, it is possible to use metal salts provided the salts are soluble in alcohol. Typically, TMS or TEOS is used as the prime constituent and alcoholic salts added to this. This is particularly useful for Group I and II elements which form insoluble alkoxides. The salts used are:

- citrates
- acetates
- formates
- tartarates
- nitrates

Chlorides and sulphates tend to be more stable than nitrates and so the anion is difficult to remove. Whilst nitrates are possibly the most popular additions a word of caution should be given here that nitrates being highly oxidizing can lead to explosions during drying if dissolved in alcoholic solutions.

Acetates are a safer alternative and often more soluble than nitrates but these materials do not thermally degrade as well as nitrates so their removal from the gel is a bit more tricky. Also many acetates are basic in solution which can lead to very rapid gelation of silicates but this can be retarded to some extent by buffering the solution with acetic acid. In the case of sodium, the acid tartarate is a good form to use since this salt already contains alcohol groups which react with the alkoxide solution. Acetate groups can be reduced in the system by reacting the acetate with the alkoxides liberating an alkyl acetate, ROAc. This reaction is achieved by heating the alkoxide and acetate together in the absence of solvent and distilling off the ester $R^1CO.COR$. In order to achieve this it is preferable to have the alkoxide to acetate mole ratio greater than one which yields a soluble hydrolysable product which is similar to a double alkoxide.

Esters of carboxylic acid $X-R^1CO.COR$ hydrolyse by a variety of mechanisms depending on the ester and the conditions of hydrolysis. There are two main reactions, basic and acidic. Basic hydroxylation takes place in pH7 upwards and acidic hydrolysis in acid solutions. This reaction consists of cleavage of the ester molecule with addition of a proton H^+ to one of the resulting fragments and a hydroxyl to the other. The two types of fission of carboxylic esters is shown below

$$\text{Acyl oxygen fission} \quad R^1-\overset{\overset{\displaystyle O}{\|}}{C}|O-R$$

bond breaks here

$$\text{Alkyl oxygen fission} \quad R^1-\overset{\overset{\displaystyle O}{\|}}{C}-O|R$$

bond breaks here

In basic solutions most carboxylic esters hydrolyse by acyl-oxygen fission but alkyl-oxygen fission can also take place under basic conditions. In acyl fission, the rate of reaction is controlled by the rate of addition of OH^-; whereas, in alkyl fission the ionization of the ester controls the rate. Again both types of fission operate in acid catalysed hydrolysis. Here the acyl-oxygen fission usually takes place in simple esters such as methyl or ethyl; whereas, alkyl-oxygen fission is more common with esters derived from tertiary alcohols, e.g. t-butyl acetate. Some acetate-alkoxide pairs that can be produced in this way are listed below:

- magnesium acetate - aluminium alkoxide
- calcium acetate - aluminium alkoxide
- lead acetate - silicon alkoxide
- lead acetate - titanium alkoxide

Sol-Gel Transition in Alkoxides

This gives a good synthesis route to lead silicate glasses in particular.

Rare earth glasses have also been prepared by this technique and these glasses find use in dental applications where radio-opacity, due to the rare earth atom, is very important.

6.3 MULTICOMPONENT AQUO-SOLS CONVERTED TO GELS

This far we have concentrated on alkoxide derived gels and we have tended to consider aqueous systems as precursors for powder formation. However, stable gels can also be produced from aqueous systems. There are various ways of achieving this but the most obvious is the hydrolysis of certain metal salts (10). Some metal salts hydrolyse to hydroxides forming a range of polymeric species. Before examining multicomponent systems, let us take the more simple example of the titania system.

Titanium IV salts can be readily hydrolysed. A good starting material is the chloride and the method of hydrolysis is by base addition, usually ammonium hydroxide. The hydroxide can then be peptized by the addition of dilute nitric, hydrochloric or trichloroacetic acid and finally some of the water evaporated off to form the gel. This peptization only occurs at low pH value, 2.8. One multicomponent system that has been prepared by this method is calcium titanate. The preparation involves the mixing of titanium chloride with calcium nitrate then dispersing this solution in a chlorinated solvent and adding an amine R C-NH (Primene JMT) slowly which gelled the aqueous phase. The structure of this gel was such that it enabled powder formation to take place as a precursor to a densifed ceramic. If the conditions of peptization are controlled, it is possible to produce monolithic gels re-dispersible in water. These gels can be used in the preparation of fibres by processes such as spinning.

A further very interesting system has been developed by Bob Shoup of Corning (11). He has developed alkali silicate gels with pore size in the range 10-300nm. The method is to mix a colloidally dispersed silica sol with potassium silicate in the presence of an amine. This gives a rigid porous silica gel. A washing procedure was then used to remove the alkali followed by drying under the influence of microwaves and then sintering. The amine used was formamide. It is possible to cast near net shapes by this method. The advantage of this method is best understood by reading the next chapter since it is possible to produce a large pore structure which reduces capillary forces during drying of the gel. More details of the process will be given in Chapter 7.

Another application of sol-gel technology that has recently received attention is the route to materials with perovskite structures. These materials have useful electrical characteristics and find applications as capacitor materials, piezoelectric materials and also superconductors. The perovskite structure is very important since many electro phenomena are related to this structure. Perovskites are basically mixed oxide structures requiring three or more different atoms for their formation. They will be dealt with in more detail in the chapter on Applications, Recipes and Reviews.

Citrate synthesis has shown great promise in the preparation of uniformly substituted perovskite oxides and is capable of enabling the formation of amorphous gels. The preparative route usually involves the conversion of the metal nitrate to citrate using citric acid. Each precursor is then mixed. Sufficient citric acid must be used to bind the metal ions as if all the NO_3 ions were replaced. Gelation occurs by the removal of water. The first water removal is usually done rapidly in a rotary evaporator under reduced pressure which generally yields a viscous liquid. This is then transferred to a vessel with a large surface area and the water further evaporated at elevated temperature, say 70°C. These gels can then be used for whatever purpose is necessary but have often been precursors to powder formation. Baythoun and Sale (12) have prepared strontium substituted lanthanum manganate perovskites. Not a great deal is known about the structure of these amorphous gels but their investigations of these materials indicate that a metal citrate or nitrate is not present in the gel. They have proposed that the metal nitrate becomes complexed with the citrate, possibly as shown in some of the schemes below where we consider the manganese nitrate/citrate gel only.

Sale (13) has also used this technique for the preparation of superconductors:

La Ba Sr CuO and Y Ba CuO

The preparative route for the gels was very similar to that just described for the substituted lanthanum manganite. Successful high temperature superconductors with Tc=93K have been made by this method.

Another report of the 1:2:3 systems from gels is made by Sakka (14). He prepared transparent gels via acetates using ammonia to adjust the pH. The solution was evaporated at 60°C to concentrate the sol. Eventually the sol was sufficiently viscous to allow fibres to be drawn. As more solvent was lost a blue gel formed. These gel fibres were then dried in a vacuum at 90°C for 1.5 hours and were then heated treated. The 5m gel fibres bloated at 200°C and gave rough hollow fibres with poor mechanical strength. Thicker fibres (1mm) were produced and heated to 900°C at 5°C/min, held at this temperature for eight hours then cooled slowly. These fibres demonstrated superconductivity at 94K which continued partly up to 62K.

(a)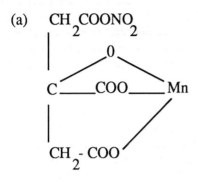

(b)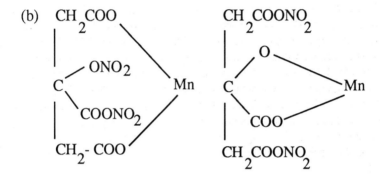

(c)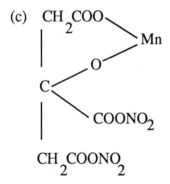

Amorphous superconductors in this system have also been produced by Dunn and Chu (15) from nitrate solutions treated with citric acid.

6.4 OTHER ROUTES TO GELS

There are several other ways that gels can be prepared without resorting to pure alkoxides, salts or aqueous systems. Certain metal oxides and hydroxides will dissolve in alcohol to form partial alkoxide solutions which can then be used in much the same way as alkoxides. Many of the oxides and hydroxides of Group I metals come into this category. Some other useful oxides are:

- lead monoxide
- boric acid/oxide
- phosphoric acid/oxide

These again can be substituted for alkoxides where these are either difficult to synthesize or sparingly soluble. Boric oxide is a common constituent of many silicate glasses and can be introduced into sol-gel systems by dissolving boric acid in methanol to form the trimethylborate. One problem with boron compounds is loss via volatilization thus the higher alcohols are used to prepare partial boron alkoxides since they reduce the risk of volatilization having lower vapour and partial pressure.

Again phosphoric acid can be used to produce alkoxides directly in alcohol and this reacts with silicon alkoxide to form phosphorosiloxane intermediates.

6.5 MIXED SYSTEMS

It is possible to produce gels from hybrid systems of alkoxides and colloids. Fine silica powder can be dispersed in hydrolysed TEOS solution with an amine present. This produces a high quality transparent silica on processing. The amine is added to adjust the pH which is optimum between 4.3 and 4.8.

Silica powder can also be dispersed in an aqueous solution of titanium isopropoxide. The fumed silica powder is peptized with acetic acid.

Other powder fillers can be introduced into alkoxides. These materials may not form an amorphous gel but may simply be suspended in a gel formed by the hydrolysis of the alkoxide. Whilst this will not give the same level of molecular homogeneity, it still gives intimate mixing when compared with the mixing and reaction of conventional ceramic powders. The gel can either act as a bond or it can react with the powder to form another phase. Alumina powder can be sintered at a comparatively low temperature if suspended in an alumina gel such as boehmite which is the monohydrate of alumina and which can be colloidally dispersed in water. The two phases can also be reacted at elevated temperature with the result that the boehmite breaks down to ultrafine grain alumina with a

Sol-Gel Transition in Alkoxides

very high surface area and this active powder appears to form a sintering bridge between the coarser grains at a temperature about 200°C lower than would be necessary to sinter the course grains to the equivalent density.

It is also possible to disperse fine alumina powder in a silica gel and bring about a reaction between the two to form pure, dense, fine grain maltite.

6.6 CHELATION

I have already mentioned the problem of mixing two or more alkoxides with different rates of hydrolysis. Rapid hydrolysis of one phase will generally lead to precipitation rather than gelation. It is possible to slow down the rate of hydrolysis of the cation in the most rapidly hydrolysable component. The way to overcome this problem is by complexing the cation of the alkoxide which hydrolyses rapidly and this can be done using a chelating agent. Before looking at the specific case of alkoxides, it will be useful to take a general look at factors affecting the stability of metal complexes particularly those of the transition metals. Throughout this section, you will encounter the term ligand. The ligand is the species which is part of the coordination shell around the metal ion and which is covalently bonded to the atom by donating its electron pair to form a co-ordinate covalent bond. Many ligands have an atom in the group which possesses an unshared pair of electrons which is called the donor centre. Each donor centre is capable of occupying a site in the co-ordination shell of the metal ion. If the ligand only has one donor centre, it is described as one-toothed or unidentate. However, it is possible for two or more donor centres to be on the same ligand in which case it is said to be bi or multidentate. Let us take the example of a typical bidentate ligand, ethylene diamine H_2N-CH_2-CH_2-NH_2. Each nitrogen atom has an unshared pair of electrons and so two molecules of ethylene diamine can form a complex around a copper atom for instance as shown on p.64.

These multidentate ligands are referred to as chelating agents. This arises from the way they enclose the central atom rather like a crab's claw (chele is Greek for crab's claw). These complexes are thus called chelates. The electrostatic field around the cations is influential in complex formation and the order of stability is influenced by the ionic radius.

If we look at period IV, we find complex stability can be ordered in terms of radius:

Order of Stability Mn^{2+} Fe^{2+} Co^{2+} Ni^{2+} Zn^{2+} Cu^{2+}

Ionic radius   91   83   82   78   74   69

```
    H₂C—CH₂              H₃N      NH₃
    ╱      ╲                ╲    ╱
  H₂N      NH₂              Cu²⁺
    ╲      ╱                ╱    ╲
     Cu²⁺                 H₃N      NH₃
    ╱      ╲
  H₃N      HN₃
```

$[Cu(NH_3)_2en]^{2+}$ $[Cu(NH_3)_4]^{2+}$

Chelated   Complexed in
                     NH_4OH

Also in the case of ions with two oxidation stages, it is usually the ion with the higher charge number that is more stable, e.g. Fe^{3+} Fe^{2+}.

The charge distribution is also important. If you imagine each donor centre donating an electron pair and each bond was true covalent bond with an equal share of these electrons between ligand and cation there would be a net charge increase on the cation of four electrons, one for each bond formed.

It is therefore useful if the cation does not have a high affinity for electrons if complexing with ligands whose donor centres are highly electronegative which has the net effect of forming a neutral position in the bond.

Mg^{2+} Ligand
Little affinity   ⊖ ⊖   Wishes to donate
for electrons   electrons
              deadlock in the
              middle

At the other extreme the noble metals which have a great affinity for electrons form the most stable complexes with ligand containing donor atoms that need to be polarized in order to donate their electrons.

Polarization opposes the Pt desire to pull the electrons away.

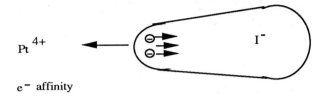

e⁻ affinity

Figure 29

There are many chelating agents. Some are very well known. Dimethylglyoxine forms a very stable insoluble red complex with nickel and is used in Ni^{2+} analysis. Perhaps the most common is EDTA, an ethylene diaminotetraacetic. This forms very stable complexes with most metal ions. It possesses six donor centres so it can coordinate up to six-fold around a cation. However, this does not often happen because of overcrowding around the central ion. The figure below shows a cobalt ion complexed with EDTA where five of the donor centres have occupied octohedral sites and the remaining sixth is occupied by a monodentate ligand (water).

To return to the subject in hand, we can use this complexing reaction to slow down the hydrolysis and condensation of metal alkoxides. Generally use is made here of ethers such as:

- methyoxy-ethanol
- isopropoxy-ethanol
- butoxy-ethanol

I can quote one example here from personal experience. On attempting to synthesize superconductors via a sol-gel route, the problem was that the alkoxides of both barium and yttrium were found to be very sensitive to hydrolysis. Barium ethoxide hydroyses if more than two parts per million of water are present. Yttrium is polymeric as the isopropopoxide and is thus not easily redissolved in isopropanol for reaction with barium ethoxide. By complexing the barium with isopropoxy ethanol, it was possible to slow down the hydrolysis rate and by adding the yttrium isopropoxide to methoxy propanol, the desire for complexation enable dissolution of the isopropoxide. Probably the methoxy-group swung around replacing the isopropoxy groups on the Y atom and effectively left a solution of the methoxide rather than isopropoxide.

7 THE GEL TO GLASS OR CERAMIC TRANSITION

The interesting stage has now arrived. We have a stable single or multi-component gel and we want to make something. The gel is a very useful starting point. From here we can:

- produce monolithic shapes
- pull fibres
- make coatings
- produce powders

All of these end points require that we dry the gel and densify it in most cases. It is this stage that often presents the most difficulties in the sol-gel process. In this chapter, we will examine these processes in some detail but there is an excellent review article by Dr. P. F. James (16) that covers the current thinking on gel structure and transformation to glass, which I can recommend for reference.

After gelation, we have a solid that is highly solvated but in which the solvent is trapped. If we dry out the solvent, there will be porosity left behind where the solvent existed and the size and amount of this porosity will further determine the nature of the resulting solid after the gel has been thermally treated in order to densify it. I have assumed, perhaps simplistically here, that we are able to actually dry the gel. Many scientists who have indulged in the practical side of sol-gel will have experienced their gels drying to a certain point then spontaneously cracking due to the high capillary forces set up in very fine pores by the included solvents. Let us therefore split up the process and consider drying first and then move onto the effects of densification.

7.1 DRYING

This is the process of removing the majority of included solvent in the gel which in the case of alkoxide derived gels will be mainly alcohol and water. As this drying stage takes place the gel shrinks by a large amount until eventually it becomes a solid which has a very high level of porosity, all be it on a microscale. Such a solid is called a XEROGEL. The usual way of carrying out this process is by very slow drying at ambient or slightly elevated temperature. The rate of drying depends very much on the size of the piece one is trying to produce. If you require a large monolithic piece, drying could take months; whereas, if you are not bothered about the gel cracking and you possibly want to mill it down further into a fine powder, the more it cracks the better, so you would dry quickly in this case. Cracking is caused by capillary forces.

It is well known that if a fine (capillary) glass tube is pushed into a liquid such as water the liquid will rise up the tube. The water will creep up the walls of the tube until an equilibrium situation is established. Because the water has an affinity for the glass as much surface area will be covered by water as possible. The only factor preventing this is the force of gravity acting on the column of water in the tube.

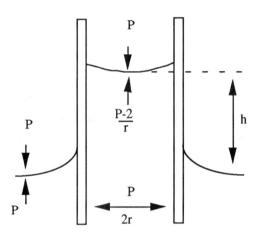

Figure 30

To quantify the situation, we have to relate the affinity that the water has for the glass by some degree of force. This face is called the surface tension. Thus, the surface tension forces balance out the hydrostatic force when the column reaches a certain critical height, h. Note that the surface of the liquid becomes curved in the tube. Despite the pressure outside being P, the pressure just below the surface is P-f(2,r) so the excess external pressure pushes the liquid up the tube until equilibrium is reached.

Pressure = $\rho \Pi r^2 hg$ for the hydrostatic column of liquid

g = gravitational content

ρ = density

Pressure = $2\gamma/r$ γ = surface tension

At equilibrium h = $f(2\gamma, \rho g r)$

The Gel to Glass or Ceramic Transition

What is significant about this equation is the fact that $h \propto f(1,r)$ so the smaller the tube diameter, the higher the liquid will rise or the higher the hydrostatic pressure. If we now consider the pores in a gel where the diameter may be nanometers then the included liquid must be at a very high hydrostatic pressure. These forces can act in such a way that they will try to collapse the pores. Thus, gels with very fine pores have a tendency to crack by this mechanism. Now there are ways of controlling this. The obvious method is slow drying which allows the gel to relieve the stresses being built up in the structure. This happens by viscoelastic relaxation. Another method is to modify the gel structure in order to increase pore size and reduce these forces. If we refer back to the effect of catalysts on gel formation, it is understandable that acid catalysed alkoxides form gels with tangled branched polymeric chains and it is the entangling of these polymer chains that form the gel. The gel structure is often quite different when base catalysed. The chains become highly branched prior to entanglement and thus begin to take on a particulate nature. In this case, gelation occurs by these species linking together. The figure below is a very simple illustration of this behaviour.

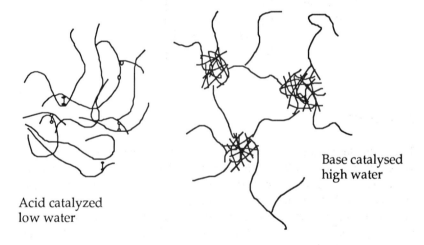

Acid catalyzed
low water

Base catalysed
high water

Figure 31

Whilst these clusters exhibit a particulate character, they should not be considered true colloidal particles as in the aquo gels such as Ludox, prepared by destabilizing aqueous sols. During drying of the polymeric gels, the structure compacts and further cross-links until the gel network is able to resist the surface tension forces so on further drying porosity will appear in the structure.

The base catalysed material behaves more like a true colloidal gel on drying. An example of the bulk density of two gels demonstrates this (17). Both have the same water content, one being acid and the other base catalysed. The acid catalysed gel had a bulk density of 1.8; whereas, the base catalysed was much lower at 0.8 suggesting that the clusters behaved as larger particles and on drying did not pack very efficiently hence the low density associated with the larger particles are larger pores. In such gels, the process of ageing is important since the clusters will continue to coalesce with time making a more rigid structure.

Emerging from this is the importance of the initial gel structure to the final solid structure that is achieved on densification and at this stage there is some level of control over the final product by subsequent control of the gelation chemistry and the rate of drying.

In a particulate gel we still have capillary forces operating and should not assume that the answer to "crack free" monoliths is merely aquo-gels or base catalysis. Imagine the gel comprised of a series of particles with all pores or interstices filled with liquid. Whilst I do not intend to prove it here, there will not be surface tension forces between liquid and solid unless there is a gas or vapour interface. As the solvents evaporate, shrinkage takes place but not at the same rate as solvent loss. This leaves pores which provide the gaseous phase, in contact with the liquid and solid.

Menisci form in the pores and the surface tension forces now try to pull the particles together.

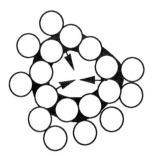

Figure 32

Also if pores of two different sizes are adjacent to each other the differential stresses generated can cause cracking. We can illustrate this on a simple capillary model.

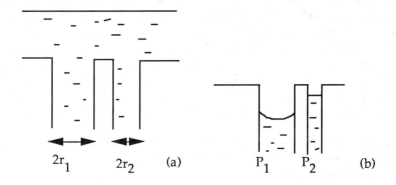

Figure 33

The differential pressure gives rise to differential stresses and if the difference exceeds to the limit for the material, cracks will appear.

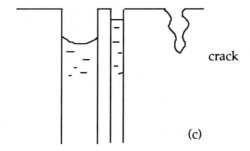

Figure 34

So we can now appreciate that the conditions which determine the final state of the dried gel are complex. To summarize there is a fundamental difference between colloidal gels and polymer gels and within the polymer gels the structure depends upon the use of an acid or base as catalyst. In general an acid cataylsed gel leads to very small pores associated with a high dry density, whereas, base catalysed gels in which the polymer molecules have formed particle clusters tend to have larger pores associated with a lower dry density. The situation is more complicated than this so you will find anomalies now and then. The complications arise by interactions with the electrolytes and solvents. Even in a "dry" state, the surface of the pores in the gel may be covered with hydroxyl or alkoxy groups. Prassas and Hench (18) have summarized this in a simple schematic diagram:

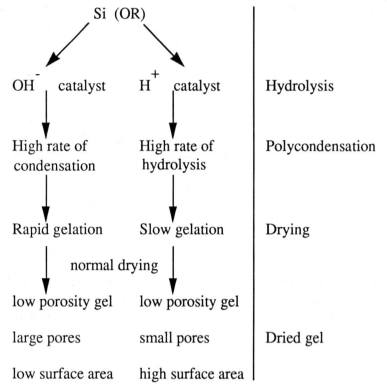

It is also useful to look at the idealized free energy/temperature curves for polymer gels, colloidal gels, glass and an ideal supercooled liquid.

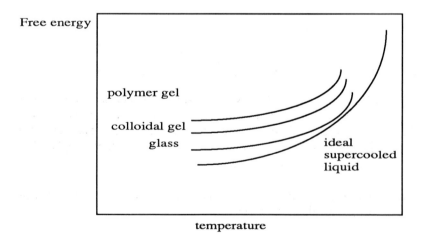

Figure 35

The Gel to Glass or Ceramic Transition

These curves represent the same oxide composition in each case. The main reason for the increasing free energy is due to the increasing surface area in the solid. Two other minor contributions come from:

- polymerization of silanol groups which is an exothermic reaction

  $Si(OH)_4 \quad SiO_2 + 2H_2O + \Delta Hf$

- The free volume of the gels is higher than the glass produced by melting since it is not possible to achieve as tight a linked network structure from the gel and during gel-glass conversion the network continues to exothermically polymerize. As you will see later, both the high surface free energy and free volume can be used to maximum effect in densification by viscous sintering enabling much lower sintering temperatures than normal to be used.

7.2 DRYING CONTROL CHEMICAL ADDITIVES (DCCA)

Once the gel has formed with thermal drying, one only has control over the drying rate. Hench (19) has pioneered the use of drying control chemical additives in alkoxide sols in order to control the drying of the gel. Specifically, these additives influence:

- the vapour pressure of the solvent in the pores
- the pore size distribution
- the drying stresses

The DCCA's that appear effective include

- formamide (NH_2CHO)
- glycerol ($C_3H_8O_8$)
- oxalic acid ($C_2H_2O_4.2H_2O$)

The investigations of these additives has not yielded any conclusive mechanisms to date but it is thought that they help produce a narrower pore size distribution thus reducing differential stresses. During gel ageing, it is also thought that the pore size increases without the width of the pore size distribution changing, thus the gel density and strength increases.

A range of glasses have been prepared using DCCA's and dried over a period of days rather than weeks. These include:

- SiO_2
- $Li_2O.SiO_2$
- $Na_2O.SiO_2$
- $Na_2O.B_2O_3.SiO_2$

Apart from the effect on pore size distribution, the formamide additive is thought to have an effect on chemical reactions taking place during the late stages of drying. If we consider silica produced from TMOS in a methanol solution then the remaining formamide and methanol in pores oxidizes in the presence of silanol and siloxane groups to form:

- formaldehyde
- formic acid
- formates

It is important that these oxidation products are removed without creating high stresses if large crack free monoliths are to be obtained. Differential Scanning Calorimetry (19) has been used to study the oxidation of residual solvent in gels with and without formamide. Those containing formamide showed much broader exothermic peaks between 250° and 300°C indicating that there was extensive oxidation of organic molecules. It was also found that at lower temperature the gels without formamide showed one endothermic peak corresponding to the removal of water, whereas, in the formamide/gel there were two peaks but the dehydration peak was at a lower temperature than the gel without formamide. (The second peaks with the dissociation of formamide.) It is proposed that formamide stops the water molecules in the gel pores adsorbing on the silanol groups inside the pores. Since the water is present as free water evaporation takes place close to the boiling point of water. Without formamide the water is adsorbed into the silanol groups via hydrogen bonds and so needs a higher temperature to desorb.

7.3 HYPERCRITICAL DRYING

Let us go back to a simple consideration of surface tension. That of a droplet of liquid on a solid surface with some degree of wetting between the liquid and solid and the liquid/solid interface in the presence of a gas and/or the vapour from the liquid.

I have already mentioned that a vapour is necessary at this solid/liquid/vapour interface in order to generate surface tension. What then if we dispense with the liquid phase? We should eliminate surface tension forces completely. Every

The Gel to Glass or Ceramic Transition

system exhibiting phase changes between solid/liquid and vapour has a triple point on a phase diagram where at a certain temperature and pressure all three phases can co-exist.

If we just move into a position where there is no liquid/vapour phase boundary then these conditions of temperature and pressure are said to be hyper or super-critical. We now have all the liquid in a vapour phase in contact with the solid gel. All we now need to do is to remove the liquid. This sounds easy in theory but requires considerable experimental skill to achieve. The basic piece of equipment is an autoclave (pressure vessel) which enables elevated temperature and pressure to be achieved. The liquid can then be removed by evacuation or flushing provided the conditions of necessary temperature and pressure are maintained. In some cases we can replace the liquid alcohol with liquid CO_2 then dry this off more easily. You can now use your imagination as to what is happening to the macrostructure of the gel. When we had capillary forces operating during drying, these forces were sufficient to cause the gel to shrink. Now we have totally dispensed with these shrinking forces, the dried gel should have almost the same volume as the wet gel. This does actually happen. The gels are referred to as aerogels as opposed to the xerogels of normal drying. These gels as solids have remarkably low density and very high specific surface area. It is possible to fully dry a gel by this process in a matter of hours rather than days or weeks and to produce large monoliths. Whereas, we considered the surface area of xerogels to be high at $600 m^2/cc$ we can produce aerogels in excess of $3000 m^2/cc$. The dried gel tends to still retain a large amount of adsorbed organic radicals but these can be driven off by heat treatment. Aerogels have been converted into clear glasses.

An additional problem arises from adsorbed OH groups but it has been found that these gels can be treated at elevated temperature in chlorine gas which is an excellent way of removing hydroxyl ions. The usual autoclave temperature and pressure range for the alkoxide process is 270-300°C and 11-14 MPa, variations on this process exist using liquid CO_2 for instance as a substitute for alcohol in the alcogel enabling temperatures of 40°C and 9 MPa pressure to be used. It is debatable as to the commercial viability of such a process in volume production.

However, there has been some interest shown in Aerogels as cavity filling in insulated cladding units for the outside of buildings. The thermal properties of these materials are quite peculiar and it is possible to achieve almost a one way heat transfer but the physics of this is not well understood at present.

7.4 DENSIFICATION AND SINTERING

Having reached the stage of obtaining a monolithic or comminuted dried gel, we now move on to the final stage which is densification. We have a material that despite drying still contains residual water mainly as adsorbed hydroxyl, also residual organics either adsorbed or chemically included. To prepare the true inorganic system, we need to drive these materials off whilst the gel is still porous and porosity is open and partially interconnected. If we try to drive off volatiles inside of closed porosity this leads to gel bloating which is a swelling and distortion of the densified gel. The conditions of time and temperature will depend very much on the original gel structure. In general, a feature of sol-gel processing is the fact that this densification takes place at a much lower temperature than one would require to make the equivalent material by a conventional route:

- fusion of oxides to form a glass
- sintering of powders to form a ceramic

The driving force for this low temperature densification is often referred to as the gel's desire to reduce surface free energy and therefore surface area but the situation is more complicated than this since this physical treatment assumes that the solid is chemically stable and in equilibrium with its surroundings. The varying degree of adsorbtion and chemical binding of each hydroxyl or organic group will dictate certain behaviour. Let us imagine the effect of increasing temperature on the gel. There will be two tendencies both competing. Firstly, hydroxyl and alkoxide groups will try to undergo further reaction. This reaction could be one with a proton or free species which would result in loss of solvent. However, groups with the required chemical affinity and close to each other may further react within the gel structure increasing the degree of cross-linking and this can result in further shrinkage and densification. The picture is therefore not simple and so it is not possible to generalize mechanisms. Instead, we tend to consider what is happening in each gel based on the physico-chemical nature of that gel. Last of all if we have managed to optimize the conditions of driving off all adsorbed impurities prior to the closing of the pores, we may find that the temperature now comes into the crystallization range. If the gel is being used to produce a glass capable of surface crystallization, we obviously have a system that is highly prone to crystallization. Here is yet one more competing factor trying to oppose densification.

Let us now examine some specific gel systems and see how scientists have observed effects and hypothesized about the mechanisms of densification.

The Gel to Glass or Ceramic Transition

At this point, it is probably best to not think of drying and densification as two distinct stages in the formation of a dense solid from a highly solvated gel. There are regimes of physical and chemical activity but these regimes overlap and the process is progressive as temperature increases. One of the most rigorous studies of densification has been made by Brinker and Scherer (20). They propose that the densification of a gel is not a simplistic process of merely viscous sintering but that four mechanisms play their part in varying degrees:

I	Very low temperature	Shrinkage due to capillary contraction
II	Low temperature	Condensation polymerization increasing the number of cross-links
III	Near Tg	Structural relaxation decreasing the free volume and therefore free energy of the network
IV	Maximum temperature	Viscous sintering where materials transport helps form the final dense solid

During heating after driving off the excess loosely bound solvents and after some shrinkage occurs due to capillary collapse, the gel will continue to cross-link. As this is happening, the free volume will decrease by structural relaxation.

Perhaps a simple analogy will help at this stage. Everyone tries to board a bus in Oxford Street at Christmas time all carrying Christmas trees. At first everyone gets on and trees stick out in every direction and the bus feels crowded. By a bit of manoeuvering, people eventually all point their trees vertically and find there is suddenly plenty of elbow room. The volume they have previously inefficiently occupied now seems so large that at the next stop the conductor is able to let twenty more people plus Christmas trees onto the bus. The people have used structural relaxation to decrease their free volume! If the system has excess free volume it will not structurally relax until a temperature is reached which enables chemical bonds to break and re-establish. Since we have a structure in a strained state that wishes to relax, it is likely that such relaxation will be an exothermic irreversible reaction. Think of the passengers on the bus; once seated or comfortable they will not want to get up and cause chaos again.

This exothermic reaction has been detected in several systems using Differential Thermal Anaysis. With this change in free volume there should be an associated increase in viscosity and again scientists have measured such increases. These changes do not take place at low temperatures but rather at temperatures near Tg, the glass transition temperature.

We can again refer to figure 35 which shows how the various gel derived glasses and fusion prepared glasses relate to each other via their free energy vs. temperature relationships. These curves illustrate that the final densified structures are not iso-structural and that the polymerized networks still have comparatively high free volume when compared to glasses made by fusion.

A good illustration of regimes is reported by Brinker, et al. (21) in a study of a multicomponent borosilicate glass which they were able to densify below 700°C. Three regions were identified over which three mechanistic reactions were postulated:

		Regime
	Temperature °C	Mechanism of Densification
I	25 - 150	Weight loss with negligible shrinkage due to the evolution of large amounts of water and alcohol but with little gel shrinkage other than that occurring as a result of capillary forces.

II	150 - 525	In this regime, polycondensation occurs and there is considerable weight loss and shrinkage due to polycondensation loss of water and from the oxidation of the organic residual compound. Shrinkage in this temperature range appeared to be from the densification of the network due to cross-links and polymerization.
III	525 - 700	Large shrinkage with little weight loss. At this stage, the rapid shrinkage is occurring by viscous flow sintering at which point the pores in the gel collapse.

The following percentages of densification are attributed to each regime:

- I: 3 percent
- II: 32 percent
- III: 62 percent

In regime III, the actual structure of the polymeric gel is important prior to drying and sintering and the apparent viscosity of the gel is a function of the degree of cross-linking of the gel.

It would appear in most gels that residual water and organics as OH and OR groups are present at very high temperatures because they are bonded as silanol groups. If surface area assists viscous sintering and the pores begin to diminish then they will eventually become disconnected and close off. If this happens before all the water and organics have been given off it can lead to swelling or bloating of the gel. It is therefore desirable to drive off as much solvent as possible. There are other methods than simple thermal decomposition. For instance, the gel can be heat treated in chlorine gas diluted in an inert carrier such as helium. This gas will react to form HCl and is therefore a dessicant gas. One problem is that Cl^- can be retained in the gel and eventually form Cl_2 which leads to the formation of gas bubbles in the gel. More recently fluorine has been introduced into gels with more effective results with regard to bubble formation. Fluorine can be introduced into the gel in several ways. Firstly, the gel can be doped with F from hydrofluoric acid or it can appear in the gaseous form by the thermal decomposition of NH_4F during the heat treatment of the gel. The Si-F bond is very strong and replaces Si-C. There is no tendency for fluorine to form gas molecules in the gel. Thus, the presence of fluorine has a very beneficial effect in reducing bubbles in the gel.

Even without the use of Cl_2 as a dessicant, fluorine can be effective in preventing swelling by being incorporated in the gel via HF. The F^- again replaces the OH^- in the silanol group.

At elevated temperatures, gel glasses may also crystallize. Unless this is carried out under controlled conditions in order to produce a glass-ceramic, it can be an undesirable effect. This needs particular consideration in multicomponent gels bearing in mind that the large specific surface area in the gel can provide nucleation sites for surface crystallization.

8 APPLICATION, RECIPES AND REVIEWS

In this chapter, I will attempt to review some of the literature which covers practical applications of sol-gel and also some recipes for a wide range of sol-gel derived glasses and ceramic powders.

I cannot be responsible for critically reviewing this work but to the best of my knowledge the majority of compositions are likely to be of practical use. These recipes will hopefully give food for thought in developing your own compositions and solving your specific problems.

8.1 SOL GEL COATINGS

Let us begin by looking at the field of coatings achieved by sol-gel techniques. This area has received the most attention in recent years and shows the prospect of successful commercial exploitation in many technologies. Coating is generally achieved via alkoxide gels but coatings have been reported in colloidal gel systems and these will be mentioned.

The range of substrates that can be coated is not limited to capability of withstanding a high firing temperature because of the associated low temperature for processing the gel. Thus, it is possible to put an inorganic sol-gel coating on a plastic. The majority of coatings have been applied by this technique in order to achieve some specific optical quality. As mentioned in Chapter I, coating from solutions is not a recent process and the literature cites examples of more than 45 years ago. Much of the progress leading to commercial coatings has been pioneered by Schott in Germany. Their process for coating architectural glass is simple. Large panes (4 x 3m) are chemically cleaned then dipped into a bath containing a hydrolysable metal compound. The pane is slowly withdrawn into a moist atmosphere which hydrolyses the film forming a transparent metal oxide layer on the surface.

A simple relationship shows that the thickness of the film is proportional to the drawing speed to a power of 2/3.

$$d \times V^{2/3} \qquad d = \text{thickness}$$
$$V = \text{drawing speed}$$

The process is that the film flows down the plate after wetting the surface. Because it wets the surface, it partly adheres and when the film reacts with the moisture in the atmosphere it undergoes hydrolysis and condensation. As mentioned previously, this film can be densified by a thermal treatment.

One major problem that arises here is anisotropic shrinkage. If the gel film attempts to shrink, it is prevented from doing so in the lateral direction by adhesion to the substrate.

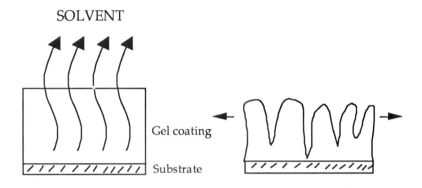

Figure 36

One distinct advantage of this technique over others is the homogeneity and uniformity of films that can be obtained. If you examine their coatings whose thickness is in the interference range with visible light any irregularities become very obvious. Schott produce window panes known as Calorex and these show excellent uniformity of coating. The technique is very versatile and allows multilayer coatings to be produced which enhances the range of optical properties that can be produced. Glasses produced from alkoxides can also be doped with transition metals such as Co, Ni, Cr and Cu. Coatings have been developed for:

- anti-reflective purposes such as coating the surface of silicon solar cells
- partially reflective coatings
- filters for ultraviolet and colour cut off
- oxidation resistance on metals such as iron and brass
- protection of moisture sensitive optical glasses
- strengthening of normal glass items by application of an appropriate sol-gel coating

The important parameters in film formation for optical purposes are the refractive index of the film and its adsorption characteristics. Some examples of metal oxide coatings that absorb in the U.V. and visible are listed below:

Absorbing in U.V. Range

Coating	Refractive Index	Upper Absorption Limit (nm)
Al_2O_3	1.62	250
CeO_2	2.11	400
In_2O_3	1.95	420
Sb_2O_4	1.90	340
TiO_2	2.30	380

Absorbing in Visible

Coating	Refractive Index	Transmitted Light
CoO	2.00	Brown
Cr_2O_3	2.38	Yellow-orange
Fe_2O_3		Yellow-red
U_2O	1.95	Yellow
V_2O_5	2.0	Greenish-yellow

There are several inherent advantages and disadvantages of using a solution method for coating. Firstly, it is possible to coat all over, inside and outside the component. The solution can coat everywhere uniformly which is a distinct advantage over vacuum coating but inside and outside coating may not be desirable. The capital equipment needed is not very sophisticated although the withdrawal from the solution needs to be as smooth and vibration free as possible. But nowadays this is easily achieved using high resolution stepping motors to drive frictionless lead screws and mounting the dip coater on an anti-vibration table. Commercial products that are made by the dip coat process are:

- Sun Shielding Glass (IROX) - The interference layers allow some parts of the incident light spectrum to be transmitted whilst others are reflected. The coating is a colloidal titania in which certain colloidal metal particles are incorporated. This gives a scratch resistant coating which does not change colour in terms of transmitted and reflected light.

- Heat Mirrors - High intensity lamps such as spots and projector lamps use halogen lamps which not only generate a high light intensity but also a great deal of I.R. (infra-red) and it is desirable to get rid of this heat. Filters can be made using sol-gel coatings which will transmit most of the visible light but reflect away the I.R.

- Cold Mirrors - This is the opposite to the above in that it transmits the I.R. and reflects the visible. It too can be used to remove heat.

- Anti-Reflective Coatings - As the name implies, these are coatings that reduce the amount of light reflected from a surface. The panel glasses on instruments in aircraft often reflect shafts of sunlight and this is where such coatings are helpful. This is also helpful in automotive rear view mirrors.

Anti-reflective coatings are of three types:

1   Single layer coatings where $n_c = n_g$; n_c = refractive index of coating

n_g = refractive index of glass

2   Multi-layer made from n_{c1} and n_{c2} where the two indices are different

3   Graded index where there is progressive change of index from the glass substrate to the external surface, usually air. This is achieved using a phase separating glass on the coating then etching one phase out to give varying porosity with depth.

Application, Recipes and Reviews 85

Anti-reflective coatings on silicon have been of considerable interest in solar cell technology. One great advantage is that the coating can be densified at 400°C, a temperature low enough not to cause excessive damage to the solar cell.

Further work has been done in spraying a mixture of colloidal silica and a partially hydrolysed silicon alkoxide onto glass surfaces. The alkoxide hydrolyses and condenses and helps form a bond between the colloidal silica particles as shown below:

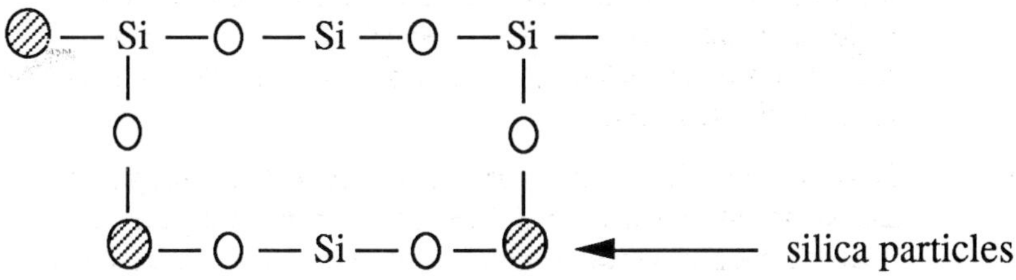

```
⊘— Si —O — Si —O — Si —
     |                |
     O                O
     |                |
    ⊘— O — Si — O —⊘   ←——— silica particles
```

Figure 37

- Colour Conversion Filters

    Interference layer coatings are another way of achieving colour filters on lamps. These coatings are applied to glasses with low thermal expansion coefficient which gives the filter thermal shock resistance.

- Laser Damage Resistant Coatings

    With the advent of higher power lasers being used in many technologies, both lenses and mirrors require special characteristics if they are not to absorb light energy in such quantity as to cause irreversible damage. This requires materials of very high purity. Many other devices can be produced for high energy laser optics including heat reflecting filters, dichroic beam splitters and polarizers. These coatings have been produced from colloidal titania and titania/silica films from colloidal gels.

- Absorptive Coatings

    These are coatings that are absorption but show no interference effects. This is achieved by incorporating finely divided platinum or palladium in a silica or alumina matrix. They give "grey" absorbtive, i.e., it is not really wavelength dependent. Less expensive alternatives have been provided using nickel oxide. They have found use in applications where glare is a problem.

Other coating methods can be employed for gels. Spin coating is a technology that has been well developed by the micro-electronics industry for the application of thin films of photoresist onto silicon wafers. This method can be applied here and consists of dropping a quantity of sol onto the substrate which is spinning at high speed, e.g. 3000 r.p.m. This gives very uniform films and is ideal for a sol which hydrolyses very rapidly since it can be kept away from moisture prior to being dropped onto the surface. As it hits the surface and spreads, it can then react very quickly with water in the atmosphere and gel. Further advantages are that it is economical in usage of solution and only one surface need be coated. Disadvantages are edge effects on non axi-symmetric substrates and the mechanical problems associated with spinning large substrates. The thickness control is a little easier than dip coating and is governed by:

- spin speed
- viscosity of the solution

A final method is spray coating. In the past, the literature has criticized this technique as the least effective of the three but more recently advances in spraying and atomizing equipment have brought this technique into line with others. In particular, it is possible to spray onto a hot surface, e.g. glass below Tg and obtain a very rapid reaction between the sol and the glass which gives coatings that develop excellent adherence and form in a very short time. The improvements to this technique have been brought about by modern techniques that enable very finely atomized jets to be produced by the use of ultrasonics or high pressure hydraulic atomisers. Very recently develoments have taken place in producing "molecular sprays".

Spraying has potential as a hot coating technique for applying coatings to glass surfaces in order to strengthen the object. Work by Schmidt at the Fraunhofer Institute and Uhlmann at M.I.T. has shown that conventional sol-gel coatings have a strengthening effect on glass. The mechanisms are far from understood but it is suggested that the coating penetrates existing flaws and that the gel forms bridges across the faces of the cracks. Uhlmann (22) has also reported nitriding the gel glass coating to increase its inherent strength. Ceramic Developments (Midlands) Ltd. has shown a strength increase on glass rods and containers coated by a hot process. A partially hydrolysed alkoxide is sprayed onto the hot glass surface between 400 and 600°C and this instantly hydrolyses and forms a coating which gives a significant strength increase. Further work is being carried out in this area.

Apart from optical properties and strength improvement, it is also possible to apply electrically conducting layers to glass by sol-gel. Gonzalez-Oliver and Kato (23) have reported work on deposition of antimony-doped tin oxide on various glass panels.

Electroconductive layers are finding applications in display panel technology, and it is suggested that they will find other applications such as in transparent ovens or could be used to heat up the glass to prevent ice or moisture build up.

The techniques of dip coating and spray coating were contrasted and in particular it was discovered that the spray coatings were more crystalline than dip coated films and that the crystals in the spray coating showed preferred orientation.

There are many reported sol-gel coatings and I have listed some of them in the table below:

Single Oxide	SiO_2 TiO_2 V_2O_5 ZrO_2 Al_2O_3 In_2O_3 various transition metal oxides RuOx RhOx ThO_2
Binary Oxides	SiO_2-GeO_2 SiO_2-TiO_2 LiNb-O_3 SiO_2-Al_2O_3 Cd-SnO_4 SiO_2-ZrO_2 SnO_2-Sb_2O_3 Ba-TiO_3 CeO_2-TiO_2 Fe_2O_3-SiO_2
Multi-component coatings	Pt-SiO_2 Pd-SiO_2 Pt-Al_2O_3 SiO_2-TiO_2-Al_2O_3 SiO_2-ZrO_2-TiO_2 SiO_2-TiO_2-(ZrO_2)-Al_2O_3 SiO_2-B_2O_3-Al_2O_3-Na_2O-K_2O Fe_2O_3-P_2O_5-CaO Fe_2O_3-P_2O_5-SiO_2

8.2 ORGANICALLY MODIFIED SOLS

It is worth mentioning the development of organically modified sols. Many polymers can now be produced with excellent optical quality and for many years plastic lenses have been widely used for opthalmic and photographic applications. The greatest drawback with plastic lenses is the fact that they very quickly scratch and this limits their life. "Scratch resistant" polymers such as polyallylethercarbonate have been developed for contact lenses but despite their reputation their resistance is nowhere close to an inorganic glass. Recently advances have been made in the development of organically modified inorganic polymers.

These coatings tend to be based on polyorganosiloxanes modified with Zr, Al or Ti. It is the inorganic backbone such as Si-O-Si that accounts for the scratch resistance yet it is possible to polymerize the system and form a dense scratch resistant coating at low temperature.

The actual system is a complex mixture containing the inorganics generally as the alkoxides, together with chemicals containing epoxide bonds which react with glycols present, methacrylates are also present. Water generated in the polymerization reaction is utilized in hydrolysis of the more readily hydrolysed alkoxides and this then controls the rate and avoids precipitation. These prepolymers have been found to be low viscosity liquids that can be further diluted in solvents such as butanol or ethyl acetate for coating applications. In effect this is a composite system with cross links in the inorganic network provided by organic groups. Thus, the network is more flexible than a true inorganic network which has the advantage of being able to accommodate stresses due to thermal expansion mismatch, and drying shrinkage of thick layers.

8.3 CERAMICS FROM SOL-GEL

Earlier in this monograph, I have stressed the importance of sol-gel in the production of ceramic powder precursors which:

- have molecular homogeneity
- have the correct particle size distribution to aid densification
- densify at temperatures below those required for conventional processing

In general the majority of sols and gels for this purpose are colloidal and aqueous or in the case of alkoxides they are base catalysed. Here the gel is only an intermediate phase so whether it holds together as a monolith or not is of no importance. In fact, it may be desirable to have a friable gel after drying which can be easily milled to a small particle size. The best method of obtaining a very fine ceramic powder, however, is by peptizing aggregated particles and then separating the peptized material from solution. Before examining some typical preparation routes in detail, it may be valuable to examine the requirements of the Ceramist with regard to the ideal form of powder needed to optimize physical properties.

8.3.1 DENSIFICATION

It is generally accepted that to produce ceramics with high mechanical strength it is necessary to form the ceramic to as dense a state as possible. Densification is brought about by sintering of powders and a ceramic powder capable of being sintered is referred to as an "active" powder. Simultaneously as densification takes place the ceramic body shrinks. The overall process takes place in a number of ways dependent on the nature of the ceramic:

- Liquid phase sintering may occur. In many conventional silicate and alumino-silicate ceramics we often add fluxes which form vitreous liquid phases at the grain boundaries of the more inert phases and help glue the crystals together.

- Reactive sintering is where some phase is formed at particle contact points either by a eutectic formation between the solid crystal phases or by the effects of doping additives in the structure which can be solid, liquid or gaseous deposits.

- Diffusional materials transport. This is diffusion at contact points between the materials.

Vapour transport merely rearranges material but does not contribute to densification. In fact, vapour transport can have the opposite effect. Take the example of two small spherical particles in contact:

The vapour pressure of the solid is not just a function of its chemical composition but also its geometry. The vapour pressure is related to radius of curvature and is higher over a positive radius of curvature than a negative one, so the material in vapour form will transport from the curved surface of each sphere into the neck regions and the coalesced particles will eventually change from A to D. (The latter stage takes a lot longer and is not a vapour transport related mechanism.) The net effect is that surface free energy has been lost but this energy is also the driving force for densification; hence, the initial driving force has been lost here without any further net shrinkage.

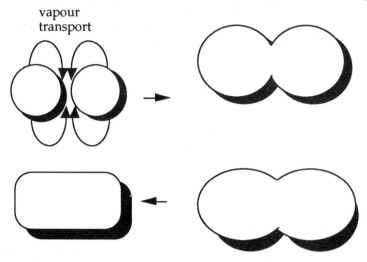

Figure 38

In the sintering process, one of the most important parameters is the particle size since the rate of densification is related to particle size as follows:

rate for viscous sintering $\quad\quad 1/r$, $\;$ r = particle size

rate for volume diffusion $\quad\quad\quad 1/r^{3/2}$

The above two relationships are at constant temperature but obviously if we reduce the particle size for a given rate we should be able to reduce the sintering temperature. The basic problem in producing powders with small particle size is that they tend to agglomerate. The agglomerated particles act as large particles in their own right even though they have a microstructure of many small particles and this limits densification.

As we have discussed earlier, it is possible to peptize agglomerated particles in order to bring powders back down to the particle size of the original structural units. Two further parameters are particle shape and size distribution. Particles with a faceted shape possess higher specific surface area than simple spheres and in fact we find certain facets more active in sintering than others due to crystallographic packing on those faces. This type of particle thus tends to have a much higher activity. Because of the difficulty of preparing fine powders in their single separated crystal form, it may not be possible to make good use of this phenomenon.

Application, Recipes and Reviews

Control of particle size distribution is, however, in the realm of the Ceramist. If we consider a monodisperse powder fairly efficiently packed together, we can appreciate that there will always be holes in the structure:

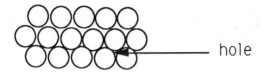

Figure 39

If the powder consists of two sizes with the smaller particles able to just fit into these holes this should lead to a denser compact:

Figure 40

Bimodal powder size distributions are often used since it is possible to get the highest green strength (unsintered) with minimum drying shrinkage.

Sol-gel technology has helped in providing routes particularly to small unagglomerated powder particles. Naturally the engineering ceramics have received much attention, particularly alumina. It is possible to produce alumina by a sol-gel route via both aqueous and alkoxide routes. Work was reported by Yoldas (24) as early as 1975 who actually produced monolithic alumina gel and sintered it. The only drawback with a conventional monolithic gel method is the excessive shrinkage that takes place. More recently, workers have been interested in the sol-gel route for providing a suitable matrix in which reinforcing fibres can be dispersed. Again shrinkage presents a serious problem since during densification shrinkage can damage fibres. A recent advancement by M. Chen

(25), et al. at Sheffield University has enabled this shrinkage to be reduced dramatically. They describe a preparative route whereby a Boehmite (alumina monohydrate) gel is prepared and mixed with a fine alumina powder.

The method involved:

    Deflocculation of alumina filler ultrasonically in pH2 with HNO_3

    Boehmite gel powder is then peptized in the above solution to give a good sol

    Sol was gelled by addition of a little aluminium nitrate

    Gel dried slowly at 50°C

    Dried gel then fired at 500°C then at 1000°C

This gave an alumina body with a comparatively low shrinkage into which they were able to incorporate carbon and silicon carbide fibres giving composites with useful strength and improved fracture toughness.

Not only are single oxides of interest but also binary and multicomponent systems. Much has been said about zirconia toughening and this has lead to many workers trying to incorporate ZrO_2 into various matrices by sol-gel.

I will digress slightly again and explain basically the policy of zirconia toughening. Zirconia toughening is also known as transformation toughening. This toughening occurs in systems where a phase can undergo a stress induced martensitic transformation. This phase change absorbs energy from the stress field near the crack tip and can thus be considered a form of plasticity in the ceramic. Zirconia can exist in three different phases. On cooling from the melt, cubic zirconia first nucleates but then transforms into the tetragonal form at 2370°C. On further cooling the martensitic changes take place to the low symmetry monoclinic form and a large volume change takes place: approximately three percent.

In order to stabilize cubic or tetragonal down to room temperature, additions of the following can be made:

-     CaO   fully stabilized cubic
-     MgO   partially stabilized tetragonal
-     Y_2O_3 tetragonal zirconia

Very fine zirconia can be precipitated in other ceramic bodies and if a crack attempts to travel past the precipitate particle the stress field can create a microcrack which can deflect the path of the original crack or "blunt it".

Hideyuki Yoshimatsu, et al. (26) report the preparation of ZrO_2-Al_2O_3 ceramics by sol-gel. Their route was via Zr-Al compounds with large organic functional groups $(CH_2)_{12}CH_3$, $(CH_2)_4COOH$ and $(CH_2)_2NH_2$. They prepared powders by gelation of the sols with ten percent NH_4OH then spray dried this gel. Other sols were evaporated in a rotary evaporator under reduced pressure at 50°C. Depending on the process, particles of different sizes and morphologies were obtained. Spinel MgO-Al_2O_3 is another useful material exhibiting a high degree of hardness. These ceramics have been successfully prepared by several routes. Dislich (27) reports the synthesis of Mg-Al spinel by reacting magnesium methoxide with aluminium nec-butoxide. This forms a mixed alkoxide solution:

$$Mg(OR)_2 + 2\,Al(OR^1)_3$$

$$\begin{array}{c} OR \\ | \\ Mg \\ R^1O \diagdown \;\; RO \;\; OR^1 \diagup OR^1 \\ \diagdown | \;\;\;\;\; | \diagup \\ R^1O \diagup Al \diagdown O \diagup Al \diagdown OR^1 \\ | \\ R^1 \end{array}$$

$$R = CH_3 \quad R^1 = CH(CH_5)_2$$

Hydrolysis is brought about and eventually a gel forms which is dried and heated to an elevated temperature to carry out condensation reactions within the gel. The crystal size in the fragments of gel were as small as 10nm. This was further heated to 1150°C and crystals were now about 100nm with formation of good spinel well below the normal temperature for sintering of oxides.

An aqueous route to spinel can also be used. This consists of blending a slurry of magnesium hydroxide into a solution of aluminium hydroxychloride in water. This system then forms a gel which can be dried, fired and milled down to powder.

Finally, advances are now being made in the sol-gel synthesis of non-oxide ceramic powders. Yoshiyuki Sugahara, et al. (28), have developed a technique for the preparation of silicon carbide. The usual synthesis route consists of mixing silicon with carbon as intimately as possible and subjecting the mix to a heat treatment under conditions of controlled atmosphere. Sol-gel provides the key to very intimate mixing. The method they adopted was to use polyacrylonitrile polymer as the carbon source, methyl triethoxysilane (MTEOS) as the source of silicon. The MTEOS solution was partially hydrolysed and the polyacrylonitrile in dimethylsulphoxide, mixed in. The gelation occurred by removal of the dimethysulphoxide. This gave a glass-like solid which was transformed to silicon carbide on heating in argon.

8.4 ELECTROCERAMICS

Apart from the formation of useful powders for the production of engineering ceramics, there is an expanding market in the prepartion of electroceramics. These materials cover a wide range of electrical phenomena.

- ferro-electric crystals respond to electric fields
- ferro-magnetic crystals respond to magnetic fields
- piezo-electric crystals generate a potential across certain faces when mechanically strained, or vice versa, change shape when charged
- capacitor ceramics high dielectric constant for the production of capacitors
- pyro-electric ceramics charge produced on heating
- electro-optic ceramics an optical property is changed by the application of a charge.

Many of these ceramics are materials based upon the perovskite structure. You have already heard about this structure in connection with superconductors. It is a very important crystal structure and worth some attention here. Mixed oxide requiring three or more different atoms are generally found in one of two crystal structures:

- perovskite
- spinel

The simple structure is the perovskite and the unit cell of the structure is represented on p.9 The general formula is ABO_3. Cation A has a co-ordination number of 12 whilst B has a co-ordination number of 6. The atom A is therefore the larger and both A and the oxygen atoms form a cubic close packed structure. The table shown on p. 94 lists some typical ABO_3 compounds with the perovskite structure:

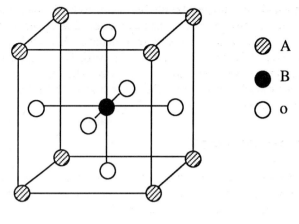

Figure 41

NaNbO$_3$	CaTiO$_3$	CaSnO$_3$	BaPrO$_3$	YAlO$_3$
KNbO$_3$	SrTiO$_3$	SrSnO$_3$	SrHfO$_3$	LaAlO$_3$
NaWO$_3$	BaTiO$_3$	BaSnO$_3$	BaHfO$_3$	LaCrO$_3$
	CdTiO$_3$	CaCeO$_3$	BaThO$_3$	LaMnO$_3$
	PbTiO$_3$	SrCeO$_3$		LaFeO$_3$
	CaZrO$_3$	BaCeO$_3$		
	SrZrO$_3$	CdCeO$_3$		
	BaZrO$_3$	PbCeO$_3$		
	PbZrO$_3$			

An interesting property of some of these structures is that whilst they relate to the perovskite, the basic polyhedra are distorted but the co-ordination number remains the same. If we examine the effect of temperature on BaTiO3, we find that as the temperature is lowered structural distortions take place. Basically, there are small displacements of the barium and titanium atoms relative to the oxygen framework. This leads to the development of a permanent electric dipole moment being established. This is characteristic of a ferroelectric crystal. One important factor in the preparation of electroceramics is the stoichiometry of the crystal with respect to oxygen. Oxygen defficiency can affect the electrical characteristics of the lattice. A particular example is that of the perovskite high temperature superconductors. Here it is very important to keep a near stoichiometric ratio with oxygen yet oxygen is easily lost from this lattice. To maintain stoichiometry, the ceramic is generally fired in flowing oxygen.

There are many ways to synthesize perovskites via sol-gel. Barium titanate has been made successfully by reacting barium isopropoxide with titanium ethoxide to form a gel then precipitating out the titanate. Phule, Raghavan and Risbud (29) have studied the effect of the starting point on $BaTiO_3$ synthesis.

Beginning with the $Ba(OH)_2$, titanium-propoxide is added to an aqueous solution of $Ba(OH)_2$. This produces $BaTiO_3$ + H_2O and C_3H_7OH.

The oxide BaO has also been used as a starting material to produce the alkoxide of barium by refluxing in dried ethanol. The alkoxide can also be prepared directly from the metal. The metal is refluxed with dry isopropanol to form the isopropoxide then reacted with titanium isopropoxide. The synthesis via the hydroxide was probably the most promising since it did not require the intermediate synthesis of a barium alkoxide. Barium alkoxides hydrolyse very rapidy even in the present of 2 p.p.m. of water so if this stage is avoidable it is preferred.

Two other useful electroceramics are:

- lead zirconate titanate (PZT), a piezoelectric material
- lead zirconate lanthanum titanate (PLZT), an electro-optic material

These systems have been prepared via the alkoxide route (30). The problem with this route is that lead alkoxides tend to be insoluble and unstable on storge. Instead of a simple alkoxide a better approach is the complexing of the alkoxide which is essentially chelation of the lead atom. This has been done with methoxyethanol. The salts of lead and lanthanum used as the starting point were acetates. By catalyzing the complexed solution with HNO_3, clear gels were made which could be dried at 150°C then fired in the range 400-800°C to form the ceramic. Much of this work is new and extensive studies of the resulting electroceramics have not yet been made.

It has already been mentioned that perovskite materials can also be synthesized via citrate gels.

8.5 GLASSES AND GLASS-CERAMICS VIA SOL-GEL

There has already been considerable discussion about glass formation from gels particularly based on alkoxides. The table below is not exhaustive but it lists some of the glasses that have been produced in the binary, ternary and multicomponent oxide systems.

Application, Recipes and Reviews

(a) Binary glasses prepared by gel route

$SiO_2 - Al_2O_3$

$SiO_2 - B_2O_3$

$SiO_2 - CaO$

$SiO_2 - Fe_2O_3$

$SiO_2 - GeO_2$

$SiO_2 - Li_2O$

$SiO_2 - Na_2O$

$SiO_2 - PbO$

$SiO_2 - P_2O_5$

$SiO_2 - SrO$

$SiO_2 - TiO_2$

$SiO_2 - Y_2O_3$

$SiO_2 - ZrO_2$

$B_2O_3 - Li_2O$

$GeO2 - PbO$

$P_2O_3 - Na_2O$

(b) Ternary glasses prepared by gel route

$SiO_2 - Al_2O_3 - B_2O_3$

$SiO_2 - Al_2O_3 - CaO$

$SiO_2 - Al_2O_3 - Li_2O$

$SiO_2 - Al_2O_3 - MgO$

$SiO_2 - Al_2O_3 - Na_2O$

$SiO_2 - B_2O_3 - Na_2O$

$SiO_2 - B_2O_3 - PbO$

$SiO_2 - B_2O_3 - P_2O_5$

$SiO_2 - B_2O_3 - TiO_2$

$SiO_2 - B_2O_3 - ZnO$

$SiO_2 - CaO - Na_2O$

$SiO_2 - TiO_2 - ZrO_2$

$SiO_2 - ZnO - K_2O$

$SiO_2 - ZrO_2 - Na_2O$

(c)   Multicomponent glasses prepared by gel route

$SiO_2 - Al_2O_3 - B_2O_3 - K_2O - Na_2O$

$SiO_2 - Al_2O_3 - TiO_2 - Li_2O$

$SiO_2 - Al_2O_3 - ZrO_2 - P_2O_5$

$SiO_2 - Al_2O_3 - Li_2O - Na_2O$

$SiO_2 - Al_2O_3 - (TO_2O, Cs_2O, Ag_2O)$

$SiO_2 - B_2O_3 - Al_2O_3 - Na_2O - BaO$

$SiO_2 - B_2O_3 - Na_2O - Al_2O_3$

$SiO_2 - B_2O_3 - Na_2O - V_2O_5$

$SiO_2 - La_2O_3 - Al_2O_3 - ZrO_2$

$SiO_2 - TaO_2 - BaO - ZrO_2$

Application, Recipes and Reviews

Many of these glasses have been produced as monoliths. The idea here is to achieve a near net shape object by the controlled drying and densification of the gel and on record there are some remarkably large monoliths that have been produced this way. The advent of drying control chemical additives may well bring the drying times down to a more practical level but in certain "high added value" applications such as doped glasses for lasers, this route may still be commercially viable and certainly yields the purest materials.

It is possible to dope alkoxide gels with various oxides. Gonzalez-Oliver (31) has successfully produced glasses with large quantities of Tl_2O, Cs_2O and Ag_2O. The alkoxides of SiO_2 and Al_2O_3 as $Si(OC_2H_5)_4$ and $Al(OC_4H_9)_2(C_6H_9O_3)$(Al-chelate) were reacted with $TlOC_2H_5$, and $AgClO_4$ in various alcoholic solutions and other solvents such as toluene and benzene. The catalysts used were both acidic and basic. A range of heat treatment conditions were tried including heating in oxygen. Under specific conditions well explained in their paper some very high quality glasses were obtained without bloating, bubbles or cracking. The advantages of the sol-gel technique here is the ability to produce glasses which would be extremely prone to phase separation or crystallization if produced by conventional fusion. Use can also be made of this to allow one to move close to the boundaries of the glass forming region of many systems where kinetically a glass would be unstable and difficult to make because of the very rapid quench rate that would be necessary. An example is the production of clear glass based on SrO-SiO_2 with a composition well within the liquid-liquid immiscibility region. This work is reported by Yamane and Kojima (32). The gels they prepared were densified into transparent glasses. The gels were prepared by reacting strontium nitrate with silicon tetramethoxide.

Gonzales-Oliver, James and Rawson (33) reported a preparation of SiO_2-TiO_2 glass which is based upon the alkoxide synthesis but with more controlled chemistry taking place.

To prepare the gel, they used the following chemicals as sources of titanium:

- titania tetra-isopropoxide - TA
- alkanolamine chelated isopropoxide - TB
- chelated glycol ester - TC

TA was found to hydrolyse very rapidly, whereas, the chelated forms TB and TC did not. These compounds were mixed with silica in the form of TEOS diluted in ethanol and methanol with the addition of acetylacetone and an amount of polymethylacrylic acid. Using TA care has to be taken on hydrolysing the solution and water should not be added directly; preferred hydrolysis was therefore atmospheric moisture, whereas, the solutions with chelated forms of titanium were able to accept water additions directly. The gels on drying and firing also behaved differently. Gels based on TA turned black at 300°C and only began to turn transparent after 12 hours at 950°C, whereas, gels based on TB turned black at 300°C but became transparent by 600°C. The effect of the polymethacrylic acid was to bring about this darkening and if in a large proportion, e.g. 30-40 percent solution (where the solution is a 0.2 weight/volume percent aqueous solution of PMA), it caused bloating of the gel. The reason for adding PMA was to use the gel in coating experiments.

With the developments in optical waveguides, there has been continuing interest in the fluorine doping of silica gels. The usual synthesis is based on partially hydrolysed TEOS to which fluorine is added as $Si(OC_2H_5)_3F$. It is often difficult to maintain the F content in the gel during sintering. The F content is dependent on the specific area of the gel and the sintering parameters. The idea is to produce tubes from the fluorine doped glasses which are then thermally collapsed into preforms and these preforms are then used to draw fibres. Glasses have been produced by several routes other than those based on alkoxides.

I have already mentioned a technique developed by R. Shoup of Corning whereby large monolithic shapes have been made by gelation of colloidal silica. The technique is interesting and somewhat unique. A colloidal silica sol is mixed with potassium silicate with an amide present, usually formamide. This is cast into a mould of the desired shape. What happens is that the silicate polymerizes in the presence of formamide and deposits onto the colloidal silica particles which act as nucleation sites. The ratio is approximately 90 weight percent potassium silicate to 10 percent colloidal silica to which 10 percent excess of formamide is added. As polymerization proceeds, the gel becomes solid. The tendency to stick in the mould during shrinkage is overcome by releasing it from the mould when it is sufficiently rigid. The aim is to develop a pore structure of 100nm in the gel. The alkali is eventually removed by a washing operation in both basic and acidic baths. After dealkalization, the next stage is drying which is performed in a novel way using a conventional microwave oven. The dried shape then undergoes a firing to 1000°C to remove water and silanol groups. The densification can then be achieved at higher temperatures but care has to be taken in designing a firing schedule that is high enough to avoid crystallization, yet low enough to avoid slumping. Whilst the overall shrinkage from gelation to densification is as high as 50 percent, the shrinkage is still predictable and thus allows near net shapes to be engineered. These structures can be made quite large and strain free and have potential applications such as large mirror supports for laser optics.

Application, Recipes and Reviews 101

Another report of large silica monoliths containing a particulate phase of silica is made by Toki, et al. (34) who added silica powder to a hydrolysed TEOS solution and base catalysed the resulting gel. The gel was then dried and sintered. They reported the fabrication of gel pieces as large as 520 x 360mm which sintered down to 420 x 290mm silica glass pieces.

Even aerogels produced by hypercritical drying have been densified (35) to produce glass rods. These gels have remarkably low density of $0.15 g/cm^3$. The residual water is removed by chlorination then the gel densified.

8.6 MICROSPHERES

At the beginning of the book, I pointed out that some of the early developments in sol-gel involved the production of microspheres of oxides such as silica. The classic works of Kolbe(36) and Stoeber, Fink and Bohn (37) describe the development of silica microspheres. They mentioned the value of such spheres in areas such as :

- model substances for calibration purposes

Later it was appreciated that monodisperse spheres could be useful in sintering and densification as a means of controlling grain size. Glass spheres produced by sol-gel are now considered to have a rather more futuristic use in inertial fuel confinement in nuclear fusion. I do not want to digress here into the theory of nuclear fusion, but I will differentiate between nuclear fission and fusion via a very simple explanation:

- fission: the release of nuclear energy when the isotopes, generally the elements from uranium onwards (i.e., getting heavier), split to form nuclei of lighter elements. On doing this, a quantity of energy is released.

- fusion: the process whereby two nuclei of lighter elements such as hydrogen and its isotopes are brought close enough together to form a new nucleus of a heavier atom which would be helium in the case of two hydrogen atom nuclei combining and again with the release of energy.

The first process has been used for many years as a source of heat for conversion to electrical power, but it has many drawbacks including:

- The starting materials are isotopes that are scarce and expensive to isolate.
- The by-products of fusion are usually highly radioactive and often toxic.
- The quantities of energy released are small compared to fusion.

However, fusion is not easy to bring about in a controlled way. We have all seen the uncontrolled release of fusion energy in the hydrogen bomb. This bomb in turn requires an atomic bomb (thermonuclear device) as its detonator.

Physicists have taken two main approaches to fusion. In Britain and Europe, the projects centre very much on the generation of ultra-high temperature plasma (plasma is a collection of atomic nuclei with the electrons stripped off) and confinement of this plasma in a magnetic field in a toroid for long enough to cause a fusion reaction. The work began with Zeta and has now advanced to JET, (the Joint European Taurus), which it is hoped will start to provide some good results towards the turn of this century. In the USA, there has been considerable focus on laser induced fusion. High energy pulsed lasers have the property of being able to compress matter to very high densities, e.g. 1000 times its normal solid density. Compression is what is needed if two nuclei are to be brought close enough together to fuse. The reaction is basically initiated by compression and high temperature. During this stage, a high level of radiation is produced which increases the reaction rate and enhances the speed of the reaction. The most likely fuel to date is a mixture of the two isotopes of hydrogen; deuterium and tritium. The energy from this reaction is in the form of high energy neutrons (14.1 MeV) which can then be used as energy in a heat transfer medium such as lithium metal or in radionuclear chemical reactions to produce gaseous fuels such as hydrogen or methane.

The basic compression mechanism is one of implosion (collapsing inwards, an effect that has occasionally been seen in evacuated television tubes). The problem then is how to confine small controlled quantities of fuel in the laser beam long enough to undergo the complete reaction. For many years, glass shells or spheres have been considered as one of the best forms of confinement of the fuel. The principle is that the laser beam hits the outer surface of the sphere (which is a shell) for a very short period. Inside the shell is the fuel. Thermal effects cause ablation of the surface of the shell and a contraction of the inner surface causing compression. The shell, thus, momentarily becomes a pressure vessel and so it needs to be almost perfectly spherical. The requirements are stringent:

Application, Recipes and Reviews

- the diameter must lie between 100 and 500 μm
- wall thickness of 0.5 to 10 μm
- concentricity of inner and outer surface better than 5 percent
- asphericity less than 1 percent
- surface irregularities less than 0.5 μm

There are traditional processes for producing such glass spheres but one of the more promising approaches has been via sol-gel. This involves the incorporation of a blowing agent in the sol prior to gelation. The solution is then gelled, dried and finally crushed to a powder of specified particle size. This powder is then introduced into the top of a heated vertical tube furnace and as the particles hit the hot zone the blowing agent expands the particles to form the microspheres. The spheres can then be filled with either a gas or liquid by permeating the vapour into the sphere at elevated temperature which then condenses on cooling. The glass must, therefore, be permeable at elevated temperature but impermeable at room temperature.

8.7 SOL-GEL FABRICATION OF FIBRES

Because it is possible to progress from sol to gel in a controlled manner, it is also possible to control the sol viscosity. This is extremely useful in the production of fibres. The expression is that solutions must exhibit spinnability in order to form fibres. The viscosity of the sol increases with time as the hydrolysis polycondensation reactions proceed. When the solution reaches about 10 poise it becomes sticky and is considered spinnable at this stage. Fibres can then be drawn from the solution. One fundamental problem is that the hydrolysis continues and the gel viscosity increases with time so you may not get other than a short period over which fibres can be drawn.

It is now believed that it is essential that a solution suitable for fibre drawing should contain linear polymer molecules. The molecular morphology depends on:

- composition of sol
- water:alkoxide ratio

If we consider the silica system, one finds that the time needed to hydrolyse TEOS to the viscosity range suitable for fibre drawing is very long. Sakka (37) discovered that the time to reach spinnability could be drastically reduced by forcibly introducing water vapour and CO_2 into the sol. These gel fibres are then converted to glass fibres in the usual way by heating. One problem at this stage is the removal of organics and early silica fibres were often black due to included carbon. As would be expected, the acid catalysed gels produce fibres but base catalysis of alkoxides generally leads to a particulate structure not suitable for fibre drawing. Some interesting points arise from Sakka's (38) work. He drew silica fibres from TEOS catalysed with HCl. His water:alkoxide ratio was kept in the range 1.7 to 2.0. He discovered that the composition of the starting solution had an effect on the geometrical cross section of the fibre whilst the viscosity of the partially hydroysed sol influences the fibre diameter. Sakka was able to produce a wide range of fibres listed below:

- TiO_2-SiO_2 (10-50 w/o TiO_2)
- Al_2O_3-SiO_2 (10-30 w/o Al_2O_3)
- ZrO_2-SiO_2 (10-33 w/o ZrO_2)
- Na_2-ZrO_2-SiO_2 (25 w/o ZrO_2)

Some useful practical hints emerged from work carried out by Mukherjee (39). The following parameters are important:

- The chemistry of the sol determines:

    i) solution rheology

    ii) the kinetics of the sol-gel transition

    iii) pore morphology

    iv) organic residues

- The environmental conditions determine:

    i) the gelling rate

    ii) the solvent evaporation rate

    iii) the nature of the gel fibre surface

Mukherjee is in agreement with Sakka about the required gel viscosity for spinnability, 10 poise, and that the spinnability improves with decreasing water:alkoxide ratio up to 1.5. With a ratio greater than 4 or if a basic catalyist is used, the viscosity increases very sharply and develops and elastic gel which is not capable of spinnability.

Cross sectional uniformity appears to be a function of the shrinkage. This seems logical since fibres have a lot of area compared to volume. If the volume change is small a circular fibre develops. Likewise the uniformity of cross sections improved with the increase of water to alkoxide ratio and the volume decreased with increasing ratio from 1 to 4. During the formation of a fibre the solvent is lost from the skin into the core. If there is insufficient water available internally for gelation, the outer skin becomes rigid due to adsorption of internal water. This is rather like baking a Yorkshire Pudding in that the sudden formation of a skin on the batter traps the residual water and the steam formed helps the pudding rise. But we can drawn an analogy between rising and bloating which is not the desired state for the fibre. If we produce gels of higher water content, the polymeric chains become more highly branched and the net result is that the fibre dries more uniformly throughout its section.

Sakka (40) performed a systematic study of the effect of the alkyl group on spinnability. His main observations were that the rate of gelation and viscosity increase were inversely proportional to the size of the alkyl group. Thus, the optimum fibre drawing viscosity range exists longer in sols prepared from the higher alkyl groups.

Great interest is currently being shown in the sol-gel technique for the production of optical fibres. The great advantage of this technique is that a wide range of glass composition can be produced in fibre form with tightly controlled composition and good homogeneity. Compared to other techniques such as chemical vapour deposition (CVD) and plasma induced CVD, the sol-gel technique is:

- simple
- controllable
- economical

The basic silica host can easily be doped with oxides such as Germania GeO_2 up to 50 percent by weight. At present the R and D is being concentrated on the production of preforms from which optical fibres can be drawn by more conventional means.

The main idea of optical fibres is to produce a fibre which does not have a uniform refractive index from inside to out. Provided the surface has a higher refractive index, the interior will be able to transmit light efficiently without too much power loss since the rays are totally internally reflected.

One approach in the production of such fibres is to fuse a tube of a glass of refractive index, n_1, over a rod of glass of refractive index, n_2, where

$$n_1 > n_2$$

This rod is then fed into the heated zone of a graphite furnace and the duplex fibre drawn out of the bottom and onto a drum.

The usual approach to this method is via mixed alkoxides but it is possible to produce homogeneous solids from colloidal gels then sinter these in order to produce a glass capable of being drawn by a thermal process. The usual dopants for the silica rods (core) are:

- TiO_2
- GeO_2
- Y_2O_3
- P_2O_5
- B_2O_3

Whilst optical fibres have received considerable attention, there is increasing interest in the development of ceramic fibres by sol-gel for mechanical reinforcement. There are also other uses for ceramic fibres. If we consider zirconia, ZrO_2, it is well known that this ceramic shows permeability and behaves like an electrolyte in the presence of oxygen.

The production of hydrogen via water electrolysis is an important process and requires a membrane to separate the electrolysis products, i.e., H_2 and O_2, yet it must allow the ionic exchanges in the liquid phase to take place. Consideration has been given to using zirconia fibre mat in this application. Thus, there is much interest in the production of ZrO_2 from gels for such purposes.

8.8 SOL-GEL IN THE PRODUCTION OF REFRACTORIES

The sol-gel process has been applied successfully for many years in the production of refractories. At first it is a little difficult to identify the technology as sol-gel since essentially the gel is used as a bonding agent to a refractory aggregate but all of the principles of condensation/polymerization still apply to alkoxides as do acid and base catalysis.

Also there are parallel colloidal aqueous gel systems used again to bind refractory phases together. The choice of binder and refractory phase depends entirely upon the application for the refractory and its shape.

This field is so extensive that I cannot deal with it in detail here so I will concentrate upon mechanisms then give a general review of where this technology is applied. Concentrating first on the aqueous systems, we find that several metals have been employed as gel binders including:

- aluminium (41)
- chromium (42)

- titanium (43)
- zirconium (44)

In the case of alumina, a polymeric gel can be prepared from the hydrated chloride

$$[AlCl_x(OH)_{3-x}]_n$$

where x is in the range 0.4 to 0.8.

The formation takes place by dissolving the aluminium chloride in an inert solvent, e.g., dialkyl ether then adding water to bring about the hydrolysis. This polymer can be mixed with an inert refractory powder such as molochite, chromate, spinel, dolomite, zircon, mullite and many others. High strength refractories have been made with this binder at temperatures as low as 1200°C.

The same technique can be applied to produce titania gels. Zirconium salts can be polymerized into gels. The starting point is a salt such as zirconium acetate and the salts can be gelled using diethanolamine or triethanolamine and water.

This binder has been mixed with many refractory powders. Another variation is a magnesia/zirconia gel mix. This system has been used to produce refractory shapes by flowing the slurry into moulds and using vibration to enhance compaction

Nitrate sulphate and chloride complexes of chromium can also be used to produce amorphous polymeric gels for binders. The gels are base catalysed and the gelling time is dependent on the type of anion and degree of basicity.

Phosphates are also used widely as binding phases. The chemistry is complex but it is thought that reactions take place at the surface of the aggregates where the phosphate monomers begin to polymerize.

Another common binder which can be gelled by acidifying is sodium silicate. I have already described Shoup's (11) work which involves the use of potassium silicate in the process of forming gels and high purity silica structures. In refractory technology, the alkaline sodium silicate is acidified using:

- carbon dioxide gas
- diacetin (45)
- triacetin

The esters di and triacetin hydrolyse generating acid. Typical esters are glycerol diacetate, ethylene glycol diacetate or glycerol triacetate.

This system has been used considerably for bonding sand in foundry moulds. Typical sands are Olivine, Chromite or Zircon if casting steel.

Aqueous silica sols are also used extensively (46) in many areas of refractory technology. Often this is achieved by controlling the pH of the sol but it can also be achieved by changing the sol concentration. A common application for aqueous silica sols is in the fabrication of ceramic shell moulds for the "lost wax" or investment casting industry. The wax cores are dipped in a slurry of aggregate such as zircon sand and the colloidal sol.

Alkoxides also feature strongly as a means of binding aggregates. The industry uses a commercial grade of ethyl silicate which contains 40 percent SiO_2 as opposed to 28 percent for normal TEOS.

8.9 SOL-GEL FOR CATALYSIS

I have expounded at great length upon the fact that gels always possess high degrees of porosity and thus have very high surface areas. It should be possible to make use of this property in catalysis and membrane technology. Again this field is still in a very experimental state but some developments hold exciting prospects for catalyst supports and membrane technology. The most attention currently seems to focus on fibres as catalyst supports. Sakka (47) reports some very interesting work involving the unidirectional freezing of fibres. This process allows a reasonable quantity of fibre to be grown at one time and in principle is quite simple but not new since the principle was reported (48) in 1980 by Mahler and Bechtold who prepared silica fibres. The principle has now been applied to the preparation of zirconia, titania and alumina.

Sakka shows a schematic diagram of the technique (p.108).

and describes some of the experimental conditions. The first main stage is that of producing hard hydrogels, the best system for fibres, ions such as Na^+, K^+, and Cl^- have to be removed. This is achieved by dialysis of the gel through a cellulose tube immersed in distilled water. The tube containing a concentrated hydrogel is lowered into a cold dry ice/ethanol mixture -78°C at a rate of 2-10cm/h. Ice crystals grow in a cellular manner in the gel as needles. The physics can be understood from unidirectional freezing and zone refining in metals.

If the lowering rate is too slow, the oxide segregates to the growth front of the ice crystals so fibres of the oxide gel do not form. If the lowering rate is higher

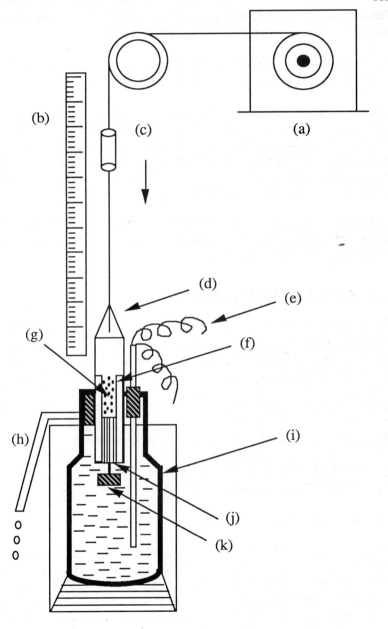

Figure 42 Apparatus used for unidirectional freezing of gel. (a) Motor, (b) Scale, (c) Guide, (d) Polyethylene cylinder, (e) CA thermocouple, (f) Polyurethane foam, (g) Gel, (h) Overflow, (i) Dry ice/Ethanol cold bath, (j) Brass bottom, (k) Weight.

then the segregation of oxide gel takes place between growing ice fibres. However, too high a lowering rate causes a large supercooling and ice now forms as discrete crystals not fibres.

Some typical process conditions are given in the table below:

	Al2O3	SiO2	ZrO2	TiO2
Precursor	Al metal	Sodium silicate	ZrOCl2.8H2O	TiCl4
Sol formation Reaction		Ion exchange	hydrolysis	hydrolysis with acid
Gel formation	dialysis	Base catalysis	dialysis	dialysis
Lowering rate	6-9cm/h	0.3-150cm/h	4-6cm/h	2.2-4.5cm/h
Fibre diameter	60-200μm	50μm	10-50μm	20-100μm
Surface area		900m²/g	350m²/g	240m²/g

The fibres are polygonal in cross section, usually hexagonal. Fibres produced this way have very high porosity. It is suggested that such fibres could be used

Application, Recipes and Reviews

as catalyst supports for enzymes. There are processes such as the conversion of milk to yoghurt where the enzymes or bacteria are lost in the process since they remain in the finished product. It may be possible for the liquid to be pumped through a catalytic column and converted on exit from the column with the catalyst rentained in the support medium, this would create a more efficient and economical process route.

Sowman (49) has reported the leaching of Al_2O_3-B_{23}-SiO_2 which leave the fibres with a uniform microporous layer around the non-porous core. HF is usually the leaching acid. The idea Sowman suggests is infiltration of this microporous layer with the appropriate catalyst. Fibres are also a good starting point for membranes because they can be woven into cloth or filament wound as tubes.

A similar field to catalyst supports is that of filtration or membrane technology for separation. Membranes can be constructed not only to separate solids from liquids but also particles from gases or gases from gases. The separation process is very dependent on the size of the pores in the membrane. The areas that sol-gel may address are mainly microfiltration which is considered to be in the range 50nm to 1μm. Smaller than this we tend to think of as separation due not to filtration but reverse osmosis or dialysis. There is a high degree of overlap here with catalyst supports which may well also be a typical filter or separation membrane. Fibres again feature here in that fibre mats have been coated with silica gel. Another way to produce a membrane is to produce a film from the gel. Films have been drawn from gels. A novel way of achieving large films but avoiding the associated shrinkage stresses on drying is by casting the gel on a liquid which

- is of geater density than the gel
- has a low reactivity for the gel
- has a high surface tension to spread the gel over its surface

S-tetrabromoethane is one such liquid suitable for this purpose. Sheets can be produced up to 0.1mm thick.

To date little attention has been paid to highly porous gels produced by hypercritical drying, aerogels. Ultra low density solids can be produced by this technique with densities as low as 0.15. The solid can also be processed to have reasonable mechanical strength by incorporation of fillers.

8.10 MATERIALS FOR ELECTRONICS

In general, glasses and ceramics used in electronic applications need to be of high purity. It is obvious that sol-gel technology has the potential to satisfy this need since the starting materials can be easily purified by distillation. The work at present is still very much experimental but scientists have some firm ideas as to the possible applications.

The fields which may benefit in the future are:

- high purity ceramic powders for packaging and substrates
- glasses for passivation layers on silicon, a wide range of sealing glasses for packaging and lead through seals
- dielectrics for capacitors

Most of the research and development in this area is currently taking place in the USA. Here, I will review only a limited number of programmes.

Cordierite ($2MgO.2Al_2O_3.5SiO_2$)

As a ceramic in electronics, cordierite is a very useful material as also are the family of glass-cermics that can be developed within the system.

High purity cordierite is an excellent material for microwave applications such as substrates or chip carriers exhibiting low loss at GHz frequencies. The dielectric constant is about half that of alumina. Cordierite has also been used to build up dense circuitry by tape casting substrates, metallising the surface then co-firing a stack of these laminates together. Some of the conductors that are compatible with cordierite are Cu, Au, Pd-Ag Mo and W. Firing conventional powders requires high temperature of 1600°C and the use of inert atmosphere because the metals used are generally Mo and W. Sol-gel processing enables cordierite to be fired at 1000°C so that lower cost metallization and firing can be achieved.

Cordierite is also a low expansion material compared to alumina, thus, the matching to silicon is much better in device packaging.

Very fine powders have been made but care has to be taken about the synthesis of a three component system in order to avoid undesirable effects such as precipitation. These instabilities are related to:

- nature of precursors
- order of mixing
- ageing of the solutions before and after mixing
- the pH of the solution
- the amount of water available for hydrolysis

Application, Recipes and Reviews

One successful recipe used:

- magnesium sec-butoxide in amyl alcohol with 1 percent wt of acetic acid
- TEOS in methanol (2 mole H_2O/1 mole TEOS)
- aluminium sec-butoxide in amyl alcohol

The solutions were all aged for 48 hours prior to mixing. The magnesium alkoxide and TEOS were mixed first then aluminium sec-butoxide added. The hydrolysis is carried out very carefully by adding the alkoxide mixture dropwise to an ammonia/water solution at 70°C and throughout the reaction ammonia was bubbled through the solution. This yields a white powder which can be separated off and washed in isopropanol. The powder has a surface area of $300 m^2/g$ so it is a very reactive powder.

The thermal treatment of this powder is also very important. Up to 900°C, the powder remains amorphous but at 920°C stuffed β-quartz appears. At higher temperature, the transformation of x-cordierite occurs (1100°C). The sintered material exhibits good mechanical strength.

Sol-Gel Derived Films

Films and coatings have been an important aspect of microelectronic technology in the production of devices. Thin films can nowadays be tailored to have the most incredible properties. Apart from the obvious applications as dielectric layers, thin films can also be used for:

- Electrochromic display devices where the film changes colour when a voltage is applied. The applications here are similar to those for LED, Liquid Crystal and Plasma displays.
- Ferroelectric films such as lead zirconate titanate.
- Nickel Ferrite ferrimagnetic films.
- Thin tantalum oxide films for capacitors.

I would like to discuss this last area in a little more detail. One method of capacitor production consists of pressing and sintering tantalum metal powder in which a tantalum wire is embedded. After sintering, the compact is anodized to produce a coating of tantalum oxide as the dielectric over the whole of the surface. Because of the porous nature of the sintered metal, the capacitor has a very high specific area of dielectric coating. Thus, we have the sintered metal as one plate of the capacitor, the dielectric and now we need the second plate.

This is achieved by infiltrating the porous structure with manganese dioxide by absorbing a manganese salt then decomposing the salt thermally. This film is built up until the pores are filled then a second contact is made to the manganese dioxide using graphite and silver loaded pastes.

In the U.S.A. there have been several programmes launched to try to achieve the tantalum coating by sol-gel rather than anodizing. It is thought that the oxide layer could be grown in a more controlled way by sol-gel and therefore it should be possible to achieve a lower defect level in the oxide coating and so be able to increase the field across the capacitor.

Not only is tantalum oxide of interest in capacitor technology but also as a replacement for silica in integrated circuits because of its higher dielectric constant. A recipe for a successful coating technique is reported as follows:

Tantalum ethoxide is dissolved in dried ethanol and water added to adjust the ratio to 1:1:3. Films had to be processed to be thinner than 3000 Angstrom to prevent cracking on drying and these were densified at temperatures up to 450°C. Beyond 450°C, there is little densification.

The dielectric constant depends strongly on the thermal history that drops through a minimum when heat treated at 500°C. However, the dielectric constant compares favourably with the value obtained by anodic growth.

Ferroelectric Films

The route to these films is shown schematically opposite. The prepared solutions were coated onto silica substrates. The thermal treatment consisted of driving off the alcohol at 100°C and burning off the 2-ethylhexanoate at 300°C. The film of PZT material can then be partially densified at 700°C.

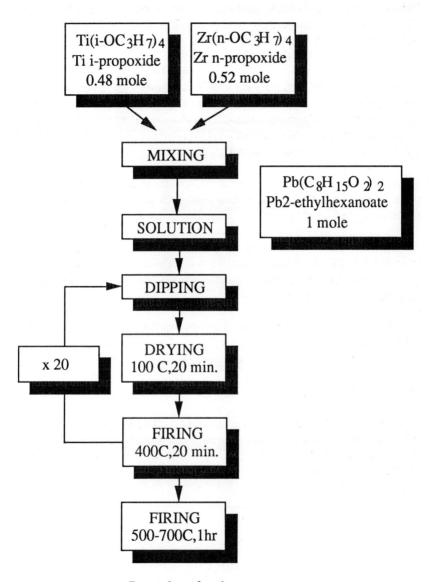

Procedure for the preparation of PZT film.

Figure 43

Ferrimagnetic Nickel Ferrite Films

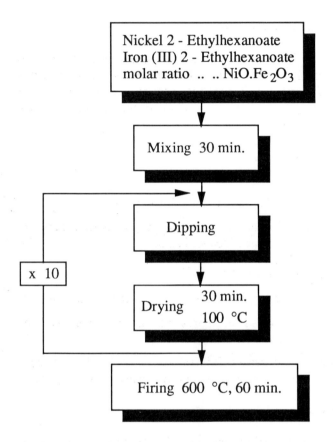

Procedure for the preparation of NiFe$_2$O$_4$ film.

Figure 44

Electrochromic Display Devices

Electrochromic displays have been made with TiO$_2$ as a fine powder coating deposited onto glass electrodes using polyvinyl alcohol as a binder. Sol-gel coatings have been developed for this application by hydrolysis of Ti(OBu)$_4$-butanol solution which is deposited onto a glass electrode by spin coating. A device was made up by placing this coated electrode in an electrochemical cell containing LiClO$_4$ in propylene carbonate as the electrolyte and using platinum as the counter electrode. The transparent TiO$_2$ film turns blue when a voltage of

Application, Recipes and Reviews

-2V is applied. On applying a reverse voltage of +2V it is possible to bleach out the colour.

Amorphous Superionic Conductors

Zirconium phosphate gels have been known for some considerable time as capable of exhibiting ionic conductivity. The general formula of the gel is $Zr(HPO_4)_2.nH_2O$. The progress of sol-gel technology has focussed investigators onto other amorphous gel systems which also exhibit ionic conduction. These include:

- lithium salts containing ORMOSILS
- Na and Li zircono-silicophosphates
- transition metal oxide gels such as V_2O_5 and WO_3

Apart from electrochromic applications mentioned above, superionic conductors can be useful in a wide range of applications as:

- cathodic materials
- low power electrochemical devices such as microbatteries or sensors
- as powdered material a solid electrolyte for applications such as the sodium sulphur battery

Sol-gel procssing shows great promise for the above technologies because of the good purity controls that can be applied and the low temperature processing.

Electronic Glasses

Electronics is now demanding very high standards of purity from raw materials used for sealing, passivation of I.C.'s, laser technology and optical waveguides.

All melting techniques present one fundamental problem and that is that a molten glass needs containment. This leads to a reaction between the glass and the container. Such reactions may have a very significant effect on the characteristics of the glass. Not only is reaction a problem but often it is difficult to achieve glass that is very homogeneous. Particularly if a glass has a high viscosity, it is very difficult to homogenize without the elaboration of stirring which enhances adverse reactions between the glass, the crucible and the stirrer or to melt at high temperatures where again the reaction is enhanced. We have already seen that sol-gel provides an excellent route for the preparation of glasses as an ultrapure scale and that good control can be exercised at the outset in obtaining very pure precursors.

Glass powders can be made ideally via a colloidal aqueous gel route, whereas, it is also possible to produce monolithic pieces via the polycondensation of metal alkoxides.

One less obvious advantage of glasses prepared by chemical route is the fact that many materials exhibit lower softening points than the glass counterpart prepared by fusion. This can be a distinct advantage in sealing operations.

8.11 BIOMATERIALS AND SOL-GEL TECHNOLOGY

In chapter 1, I gave a very brief "modern history" of sol-gel technology as applied by man to solve technological problems but if we now look at nature we can find many examples of gels and their role in biological materials synthesis and, of course, this would rank as "ancient history". It is not my intention in this chapter to review the development within life forms of biological gel derived materials but rather to look to the future and see how the field of biology could lead into technical areas through the development of new materials with physical properties far superior to materials currently available. This area has recently received attention and there are now jargon phrases such as:

- nanometer technology
- intelligent materials

which all refer to the different way that these materials must be studied and viewed.

Before looking at sol-gel related areas specifically, let us look back at the excellent materials that grow or develop naturally. One property of great importance in both nature and technology is the fracture toughness of a material.

In chapter 8.12, I will discuss a simple approach to improving the fracture toughness of inherently brittle ceramic materials by means of fibre reinforcement. This is a crude approach since we rely on macroscopic fibres embedded in a continuous matrix to create increased work of fracture via:

- fibre pull out (frictional effects)
- crack deflection (increase in fracture area)

If we look at tough biological materials, we find that their microstructure is two or three orders of magnitude finer than man made composites and considerably more complex. Sometimes this complexity is hierachichal. For instance, if we consider a hair made from keratin we find that our starting point is the α-helix molecule and these ascend as shown in the illustration on p.119.

Application, Recipes and Reviews

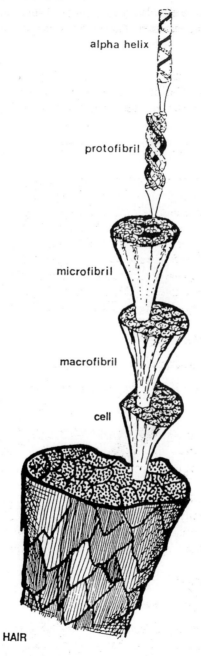

Figure 45

The actual structural changes are not important at this stage. What is worth understanding here is that nature begins its engineering five orders of magnitude down from the size of the finished product. Nature is also able to control structure within a hierachichal level independent of interactions at other levels.

Examine a tough natural material such as antler bone and we find that the work of fracture is of the order of 14,000 Jm^{-2}. Yet, the ceramic or inorganic content of this material is 60 percent. What is very different, compared to engineered materials, is the scale of microstructure. Antler bone has a "ceramic phase" of approximately 0.2µm. Monolithic ceramics exhibit work of fracture in the range 2-10 $J m^{-2}$.

Another very interesting composite is nacre which contains up to 95 percent ceramic material in a plate-like form and which makes up the structure of shells of snails and shellfish. The ceramic plates are 0.5µm thick and made from aragonite (calcium carbonate). Shell is undoubtedly a fairly tough material especially if one realizes that only 5 percent of the makeup is a polymene material. Nature has made good use of the fact that organic matrix materials cost considerable amounts of energy to synthesize and thus the nacre composite is architected to be effective with a high chalk content and very little "biological resin".

Materials found in nature can also be considered to exhibit a form of intelligence. This statement should not be taken too literally. What is meant is that a structure such as a bone may be found to have variable mechanical properties at any point dependent upon the loads it is expected to carry. Thus, at a point of low stress a less dense bone will form, whereas, in high stress regions the strongest bone material forms. The material evolves this way itself stimulated by the environment with which it interacts and aided by the D.N.A. message architecture. The material is therefore considered to possess some degree of intelligence.

For some years now small groups of biologists have been suggesting to materials scientists and technologists that they should be trying to copy natures synthesis of materials which would be much more effective in the synthesis of materials. One group in particular that has been very active in this area is at Reading University, Zoology Department where Dr. Julian Vincent and co-workers are looking at the synthesis of useful man made materials based on their natural counterparts. They are currently looking to the use of gels as synthesis media for ceramic composites. Biological gels are capable of developing the growth of calcium or silica based materials which precipitate out of the gel. The gel, however, is not merely the means by which the ceramic phases form but rather the matrix material itself eventually being incorporated as part of the structure. These gels are aqueous systems and in nature they are generally based upon

Application, Recipes and Reviews

polysaccharides. Simple organic gels in plants are gels such as:

- carrageenan (from seaweed)
- alginates (from seaweed)

The polysaccharide molecules in these gels are capable of forming helices which intertwine with each other and form structures which are very stable in water. These are stable cross-linked gels. Gels, however, found in many animal systems are less cross-linked exhibiting a greater degree of viscoelasticity and flow under shear stress. Typical of this gel structure is hyaluronic acid which is found in synovial fluid in joints or the vitreous humour of the eye.

This acid holds vast amounts of water and acts in a different way to a load bearing structure. Load bearing gels in animal systems are a mixture of polysaccharide gels, proteins and fibres; generally collagen. Interactions between polysaccharides and proteins are thought to control the size, orientation and rate of formations of collagen fibrils. The collagen fibrils give great mechanical strength to the system. Sometimes the fibrils are found in very complex orientations such as helical windings in order to support loads most effectively. This collagen/gel system is capable of growth and incorporation of ceramic phases. What happens is that calcium salts precipitate within the collagen structure. The gel controls the solubility of the calcium salts and acts as the transport medium. Collagen thus provides sites for crystal nucleation and growth in its structure. It is hypothesized that the collagen molecules provide an array of charged groups spaced in a similar way to the hydroxyapatite crystal lattice which then provides a site for epitaxial growth in the case of say bone or tooth formation.

The understanding of these principles leads to the exciting use of gels in the synthesis of man made materials. It is not essential here to copy nature directly but to make use of some of the basic principles. Vincent has proposed that simple polysaccharide gels from plants could be used as media not simply for the precipitation and growth of calcium based materials but also for the inherently stronger siliceous phases. The crystal formation in the gel could be induced by removal of solvent. There should also be techniques which can be employed to create epitaxy in the gel if needed or to actually grade the amount of the ceramic phase growing.

After precipitation, the process would be to remove solvent from the gel and induce cross-linking reactions to thus obtain a high ceramic volume fraction of submicron particles in a strong matrix. Crystal morphology may well be controllable and thus if may be possible to produce fibrous ceramic crystals and even gels which on cross-linking may themselves form an interwoven fibrous structure.

Such concepts give a very futuristic flavour to materials technology but should be easily demonstrated as feasible or not within the next five years.

8.12 COMPOSITE MATERIALS

Despite the great interest that has been shown in the improvement of the fracture toughness of ceramics through zirconia or transformation toughening, the best materials that have been achieved to date are about as tough as cast iron which still does not provide great security to the engineer who is used to designing with predictable materials.

The alternative approach to the production of tough ceramics via composite materials. For instance, the whisker or fibre reinforcement of a brittle ceramic matrix can offer the levels of fracture toughness that are acceptable for an engineering material. There is considerable research and development occurring at present using a wide variety of fabrication techniques. Many techniques, however, deposit the matrix around the fibres or disperse the whiskers in the matrix via a liquid phase which is often a fine slurry of matrix powder in a carrier liquid.

This gives sol-gel a strong attribute because of the necessity to process in the liquid phase and the possibility of using low viscosity liquids as derived from organic precursors such as alkoxides.

Unfortunately at this time there are not many fibres or whiskers which are produced in quantity at commercially attractive prices so the exercise has been somewhat academic to date, however, this situation is likely to change in the near future. In fact, there is already considerable research taking place using sol-gel technology as a fabrication technique for ceramic fibres. Also there are existing low cost fibres such as carbon fibre which normally would be fairly inert in bonding to many ceramic matrices. Whilst a weak interface is indicative that the composite will exhibit a high work of fracture, this is gained at the expense of mechanical strength and stiffness. If the fibre/matrix interface is weak, the matrix cannot transfer stress to the fibre. The art of composite design is control of the interfacial strength and therefore the ability to optimize the strength and toughness. If the surface of a carbon fibre is coated with silica or a silicate based sol-gel material this can be reacted thermally with the carbon to form an intermediate or interfacial layer of silicon carbide. The surface being silica or a silicate is now compatible with many oxide ceramic.

Application, Recipes and Reviews

The basic sol-gel approach to composite preparation consists of making maximum use of the fact that there is a liquid phase at the sol stage. This enables whiskers to be dispersed in the liquid by techniques such as ultrasonic dispersion, defloculation or high shear mixing. In the case of continuous tows of fibres these can be drawn through either a sol or a sol containing ceramic fillers as a means of depositing the matrix. After drying the compaction is made by:

- sintering the gel
- pressing and sintering
- hot pressing   } Increasing density
- hot isostatic pressing

The aim is to achieve a high density composite and in general the density increases as shown above. It should be possible to achieve high density by merely sintering the reinforced gel monolith alone but to date this has not been the reality. The physical chemistry of such systems is so complex that the practical methods have not yet been perfected. The fibres present such high surface area to the gel matrix that drying shrinkage causes a high crack and defect density in the composite. This is a major cause of low density and poor mechanical strength. A very critical stage is in the drying of the composite. It is the development of high surface tension forces during drying that lead to shrinkage cracks and geometrical distortion. The basic idea is that the composite should dry in such a way that it remains as a monolith but with a high reactivity due to the large specific area created by gel porosity. A detailed study of the drying behaviour of composites based upon Al_2O_3/SiC composites was made by Krabrill and Clarke (50). They carried out a systematic study of the effects of drying parameters on the state of the gel, and found that conventional air drying under controlled laboratory conditions lead to warping due to the fact that the relative humidity was not controlled and therefore solvent evaporation was not controlled. With controlled drying under constant temperature and humidity conditions, they were able to achieve controlled drying without cracking and at a much higher rate. Care has to be taken not to lower the relative humidity too much since this drives the drying process too fast and again leads to warping. Their final conclusion was that to produce monolithic gel composites, the humidity has to be reduced at a higher rate no greater than the transport of moisture via the capillary structure in the gel.

There will always be inherent problems when trying to produce a pure fibre/gel composite due to shrinkage and perhaps this is not an attractive commercial route. James and Chen (25) have addressed this problem by using a filler powder in the gel system. They have produced composites from SiC fibres in alumina. Typically the system is prepared by mixing fine alumina powder in a boehmite gel. The powder was deflocculated ultrasonically with nitric acid at pH2. The sol could then be gelled by adding aluminium nitrate. The fibres were incorporated into the gel by a net coating technique. The dried composite was then fired at 1200°C in nitrogen. They were able to achieve quite respectable mechanical strength of the order of 370 MPa at a fibre volume fraction of 0.5 and a reasonable work of fracture but the composite was far from optimized and still had a residual porosity at 35%.

Whiskers present a similar problem in that their incorporation effects the residual porosity of the composite. Tiegs and Becker (51) found that the final density of their composites depended on the mean length of the whiskers and they were able to improve the green strength and final density by ball milling the whiskers into their ceramic slurry. During ball milling the mean fibre length was reduced.

Densification can be approached by other methods than temperature and pressure and one such way that has been investigated by Park (52) and Lee who prepared the matrix by hydrolyzing TEOS then added three ceramic precursors:

- furfuryl alcohol (FuOH)
- chromium acetate (CAH)
- colloidal alumina (Nyacol)

The composite was prepared then by the addition of SiC whiskers. They gelled the compacts and dried them slowly at 60°C then subjected them to a range of heat treatments from 700°C to 1500°C for two hours in argon.

In the case of FuOH the carbon developed during pyrolysis reacted with the SiO_2 matrix to form SiC. The amount of SiC forming depended on temperature being greater the higher the temperature. But increasing SiC content lead to lower density.

The presence of alumina in the composite also has an interesting effect. In the presence of both Cr_2O_3 and carbon the density of the Al_2O_3-SiO_2 system is enhanced. Carbon therefore has a beneficial effect. At too high a temperature the carbon creates gaseous phases SiO(9) and CO(9) which have a detrimental effect on densification. In most cases firing at 1200°C gave composites with the maximum density and in the composite with SCA and SiC whiskers the density achieved was about 83% of theoretical.

Application, Recipes and Reviews

Apart from fibre reinforcement of ceramics, composites are also being developed with a particulate phase such as zirconia an alumina matrix. The problem in dispersing zirconia in a second phase via an alkoxide sol-gel route is lack of control over the microstructural parameters. However, colloidal gel processing shows greater promise. The basic problem in a colloidal system is instability. We have two materials of differing density, size, shape and surface charge. This can be overcome by using highly concentrated slips containing a polymer stabilizer to consolidate the system in the flocculated state. The second method which is better for binary systems is brought about by adjusting the pH of the solution before consolidation to a range where the secondary minimum of coagulation is maximized. This gives good green state density with homogeneous mixing. Dense green bodies with 10 percent (vol) ZrO_2 have been prepared by this method and fired at 1475°C for one hour to give good sintered density in excess of 97 percent of theoretical. This technique can be adapted to a range of binary composites.

Mackenzie (53) has produced several interesting composites as follows:

- polymer impregnation of an inorganic gel matrix

- porous composites consisting of a dispersed phase (solid particles in a gel)

- three phase composites obtained by polymer impregnation of porous composites

The first system was prepared by polymerizing a monomer which is impregnated into the porous gel. He has impregnated silica gels with a range of polymers including polymethyl-methacrylate, (PMMA), silicone and copolymers of PMMA and butyl-acrylate. He obtained transparent material with a range of refractive indices dependent on the polymer type and its volume fraction.

To make porous composites, an inert filler is dispersed in the sol then the system is gelled. The systems reported are based on:

- SiO_2/SiC (particles)
- SiO_2/SiC (whiskers)
- SiO_2/Al_2O_3
- SiO_2/Si_3N_4
- SiO_2/TiC
- SiO_2/Al (metal powder)

As we have already discussed in the preparation of alumina, the addition of a filler can have a beneficial effect on drying shrinkage by reducing it substantially and enabling large monolithic pieces to be produced. By firing a 33 SiC/67 SiO_2 composite gel to 600°C, a density of 2.4 was obtained.

By now infiltrating the porous composite with a polymer a three phase composite can be made which exhibits some interesting mechanical properties such as high fracture ductility, low porosity, good strength and abrasion resistance. To date there has been little said about the potential uses of such composites but no doubt applications will be found for such materials.

REFERENCES

(1) K. B. Blodgett, J. Am. Chem. Soc. **57**, 1007 (1935)

(2) W. Geffcken and E. Berger, Reichspatentamt 736411 (1943)

(3) H. Schroeder, Phys. Thin Film, **87** (1969)

(4) US Patent 3,480,458 1969 Dislich et al.

(5) US Patent 3,640,093 1972 Thomas

(6) US Patent 3,759,638 1973 Dislich et al.

(7) W. Stoeber, A. Fink and E. Bohn, J. Colloid and Interface Science, **26** (1968)

(8) D. L. Segal, J. Non-Cryst. Solids, **63** (1984)

(9) J. L. Woodhead, British Patent, 2,111,966 A (1982)

(10) J. L. Woodhead, British Patent 1,342,893 (1974)

(11) R. D. Shoup,'Colloid and Interface Science', (M. Kerker. ed.) **3** (1976)

(12) M. Baythounand F. R. Sale, J. Mat. Sci. **17** (1982)

(13) F. R. Sale and F. Mahloojchi, 12th Int. Tech. Colloquium on Ceramics Processing (Oct 87) Italy

(14) S. Sakka, Better Ceramics Through Chemistry III MRS **121** 639 (1988)

(15) B. Dunn C. T. Chu et al. Adv. Ceram Materials **2** 3B (1987)

(16) P. James. J. Non-Cryst. Solids **100** (1988)

(17) M. Nogami, Y. Moriya, Yogyo Kyokai Schi, **87** (1979)

(18) M. Prassas and L. Hench, Ultrastructure Processing of Ceramics, Glasses and Composites. J. Wiley & Sons, NY (1984)

(19) L. L. Hench, 'Glass-Current Issues' NATO ASI (1985)

(20) C. J. Brinker, G. W. Scherer and E. P. Roth, J. Non-Cryst. Solids **72** (1985) 345

(21) C. J. Brinker, et al., Ibid., **72** 369 (1985)

(22) B. Fabes, W. Doyle, L. Silverman, B. Zelenski and D. R. Uhlmann, Stronger Glass Via Sol-Gel Coating

(23) C. J. R. Gonzales-Oliver, I. Kato,J. Non-Cryst Solids **82** (1986)

(24) B. E. Yoldas, 77th Ann. Meeting Am. Ceram. Soc. May 7 (1985)

(25) M. Chen, J. E. Bailey, P. F. James and F. Jones, J. Non-Cryst Solids, **100** (1988)

(26) Hideyuki Yoshimatsu, Ibid. **100** (1988)

(27) M. Dislich, Angewandte Chemie, **10** 6 (1971)

(28) Y. Sugahora, et al. J. Non-Cryst. Solids **100** (1988)

(29) P. Phule, S. Raghavan and S. Risbud, J. Am. Ceram. Soc. **70**, 5 (1987)

(30) K. Budd, S. K. Dey and D. A. Payne, Sol-Gel Processing of PZT and PLZT Films

(31) C. R. J. Gonzales-Oliver, J. Non-Cryst. Solids **82** (1986)

(32) M. Yamane, T. Kojima, Ibid. **44** (1981)

(33) C. R. J. Gonzales-Oliver, P. James and H. Rawson, J. Non-Cryst. Solids **48** (1982)

(34) M. Toki, et al., Ibid. J. Non-Cryst. Solids. **100** (1988)

(35) Science of Ceramic Chemical Processing, ed. Hench and Ulrich. J. Wiley, NY (1986)

(36) G. Kolbe, Das komplexchemische Verhalten der Keselsaure Dissertation, Jena (1956) Ref. (d)

(37) S. Sakka, Treatise on Materials Science and Technology, **22**, Academic Press N.Y. (1982)

(38) S. Sakka, Am. Ceram. Soc. Bull. **64** (1985)

(39) Mukerjee, Phalippou 'Glass-Current Issues' Martinius Nijhoff Publ.(1985)

(40) S. Sakka, Trans. Indian Ceram. Soc. **46** (1987)

(41) U.K. 1142201 (1969) T. H. Isherwood, M. Palfreyman, R & D Polymers Ltd

(42) N. V. Arkhincheva, G. D. Nekhlanova, M. M. Sychev, Inorg. Materials (1978) **14** (6) 900

(43) U.K. 924510 (1963) D. N. Hunter, Artrite Resins Ltd

(44) U.K. 1498624 (1978) H. G. Emblem, I. R. Walters, Zirconal Processes Ltd

(45) Report from working group Pll - British Foundryman (1978) **71** 147-80

(46) R. K. Iler, The Chemistry of Silica, J. Wiley, New York (1979)

(47) S. Sakka. Sol-Gel Technology (edited by Lisa C. Klein) ISBN 0-8155-1154-X P141 (1988)

(48) W. Mahler and M. F. Bechtold Nature **285** 27-28 (1980)

(49) M. G. Sowman. Sol-Gel Technology (edited by Lisa C Klein) ISBN 0-8155-1154-X P162 (1988)

(50) R. H. Krabrill and D. E. Clark, Better Ceramics Through Chemistry II MRS **73** 641 (1986)

(51) T. N. Tiegs and P. F. Becher, J. Am. Ceram. Soc. Bull. **66** (2) (1982)

(52) S. Y. Park and B. I. Lee, J. Non. Cryst. Solids **100** 1-3 (1988)

(53) J. D. Mackenzie, et al. Materials Research Society Symposium **73**, 809